natural selection and its constraints

natural selection and its constraints

Oliver Mayo

Biometry Section, Waite Agricultural Research Institute
The University of Adelaide, Glen Osmond, South Australia

1983

 ACADEMIC PRESS

A Subsidiary of Harcourt Brace Jovanovich, Publishers
London   New York
Paris   San Diego   San Francisco   São Paulo
Sydney   Tokyo   Toronto

ACADEMIC PRESS INC. (LONDON) LTD.
24/28 Oval Road, London NW1 7DX

United States Edition published by
ACADEMIC PRESS INC.
111 Fifth Avenue, New York, New York 10003

British Library Cataloguing in Publication Data
Mayo, Oliver
  Natural selection and its constraints.
  1. Human evolution   2. Natural selection
  I. Title
  573.2   GN281

  ISBN 0-12-481450-6
  ISBN 0-12-481452-2 Pbk

      LCCCN 82-074142

    Typeset by Oxford Verbatim Limited
      Printed in Great Britain by
  T. J. Press (Padstow) Ltd., Padstow, Cornwall

preface

It is frequently stated that natural selection, in Herbert Spencer's epigrammatic phrase the "survival of the fittest", is a tautology, that because natural selection can explain anything, it is vacuous. It is claimed that the process is too slow to explain evolution in the time available, that it is too imprecise to explain the precise adaptations found in nature, and that it is negative, not positive, in that it only accounts for the elimination of deleterious traits and hence genes. It is said that natural selection cannot explain progressive evolution nor can it explain speciation. One hundred years have passed since Darwin died, a little over fifty since Fisher wrote *The Genetical Theory of Natural Selection* and fifty since Haldane wrote *The Causes of Evolution*. The time should be ripe to draw up a balance sheet for the theory of evolution by natural selection.

In this book I aim to provide answers, necessarily incomplete, to the following questions.
(1) How much evolutionary change can be explained best by natural selection?
(2) How much evolutionary change can best be explained by other mechanisms of evolution which have been advanced at various times?
(3) Under what circumstances will natural selection be completely or partly ineffective?
(4) Within what constraints must natural selection operate?
(5) Are the constraints on natural selection unique to natural selection?

For the most part, the discussion is very general, but I have examined a few cases in some detail for the light that they shed on the questions posed above.

February 1983   *O. Mayo*

In memoriam J. A. Bishop

contents

preface   v

acknowledgements   ix

1   alternative hypotheses of evolution   1
   1.1   The Problem   1
   1.2   The Corpus Callosum as a Specific Example of the
        Problem of Evolutionary Explanation   6
        Topics for Discussion   8

2   general boundaries for evolution   10
   2.1   Time   10
   2.2   Information   12
   2.3   Complexity and Order   14
        Topics for Discussion   16

3   physical constraints   17
   3.1   Overall Constraints   17
   3.2   Allometry   18
   3.3   Life History   19
   3.4   Directional Change   20
        Topics for Discussion   22

4   early evolution   24
        Topics for Discussion   26

5   adaptation   27
   5.1   Non-adaptive Directional Change   30
   5.2   Precision of the Mechanism of Natural Selection   32
        Topics for Discussion   33

6   the rate of evolution   34
   6.1   Single Gene Selection   34
   6.2   Rate of Change in Fitness   37
   6.3   Rate of Amino Acid Substitution   42
   6.4   Rate of Change in a Quantitative Trait   48
   6.5   Rate of Karyotype Evolution   52
        Topics for Discussion   58

7   canalization   59
7.1   Dosage Compensation and Duplicate Gene Inactivation   60
7.2   Clutch Size as a Canalized Trait   62
7.3   Implications of Canalization   65
     Topics for Discussion   69

8   processes regarded as distinct from natural selection   70
8.1   Sexual Selection   70
8.2   Kin Selection   71
8.3   Evolution of Senescence   74
     Topics for Discussion   76

9   sexual reproduction   77
9.1   Direct Consequences in Relation to Selection   77
9.2   Reproductive Isolation   78
     Topics for Discussion   88

10   speciation   89
10.1   Introduction   89
10.2   Types of Speciation   90
10.3   Rates of Speciation   95
10.4   Extinction   97
10.5   Patterns of Speciation and Extinction   99
     Topics for Discussion   101

11   conclusions in terms of historical science   102
     Topics for Discussion   105

references   106

author index   133

subject index   141

acknowledgements

I wish to thank a large number of people for provision of data, answers to questions, and criticism of the manuscript: B. O. Bengtsson, J. H. Bennett, B. C. Clarke, J. Felsenstein, I. R. Franklin, P. D. Gingerich, L. Godfrey, D. L. Hayman, M. A. Jeeves, C. R. Leach, M. J. Mayo, H. Measday, M. Nei, P. R. Phillips, D. A. Simpson, J. A. Sved, D. E. Symon, G. Thomson and H. R. Wallace.

I thank K. Kain and H. M. Simpson for fine typing and secretarial work.

I thank the following publishers and authors for permission to reproduce portions of the following references:

Cambridge University Press for J. R. Gold (1980) Chromosomal change and rectangular evolution in North American cyprinid fishes. *Genet. Res., Camb.* **35**, 157–164.

Systematic Zoology Editorial Offices for R. H. Brady (1979) Natural selection and the criteria by which a theory is judged. *Syst. Zool.* **28**, 600–21.

Evolution for J. B. S. Haldane (1949) Suggestions as to quantitative measurement of rates of evolution. *Evolution* **3**, 51–6.

Springer-Verlag and Dr D. H. White for D. H. White (1980) A theory for the origin of a self-replicating chemical system. I: Natural selection of the autogen from short, random oligomers. *J. Mol. Evol.* **16**, 121–47.

Elsevier/North Holland Biomedical Press for P. E. Brandham (1977) The frequency of spontaneous structural changes. In *Current Chromosome Research* 1976, (eds. K. Jones and P. E. Brandham) pages 77–87.

Dover Publications Inc. for R. A. Fisher *The Genetical Theory of Natural Selection*.

Experientia for O. Mayo (1980) Variance in clutch size. *Experientia* **36**, 1061–2.

Genetics for A. R. Templeton (1980) The theory of speciation via the founder principle. *Genetics* **94**, 1011–38.

Canadian Journal of Genetics and Cytology for S. H. Boyer, A. F. Scott, L. M. Kunkel and K. D. Smith (1978) The proportion of all point mutations which are unacceptable: an estimate based on hemoglobin amino acid and nucleotide sequences. *Can. J. Genet. Cytol.* **20**, 111–37.

Nature for J. E. Cronin, N. T. Boaz, C. B. Stringer and Y. Rak (1981) Tempo and mode in hominid evolution. *Nature, Lond.* **292**, 113–22.

1 alternative hypotheses of evolution

1.1 THE PROBLEM

Given a structure, organ, tissue or function, can we explain it in adaptive terms? Given that we can do this, are we any further forward or are we simply ingenious phrase-makers, as suggested by, for example, Lewontin (1978b) or Brady (1979)?

The answers must depend on several other contingent questions. Can we explain the case in some other way? Is this other way more general than adaptation as a result of natural selection, or is it *ad hoc*? Do we need additional, unstated hypotheses in either case? To what extent can either hypothesis be tested?

In essence, an evolutionary explanation is a historical explanation, and so may not yield predictions which can be tested now. Furthermore, when an evolutionary rule is enunciated, it is usually a statistical rule, true over long periods but not necessarily in any short-term case (Beckner, 1959). For example, Williston's rule, that nearly identical structures will either merge or specialize, is the result of collation of very extensive fossil data, but has many exceptions and cannot be used for prediction in a single case, except in a statistical sense. (This is also true of Mendel's laws, but is rarely regarded as discrediting them.)

Because of the general difficulty of short-term prediction and the specific difficulty of historical prediction and retrodiction, it has frequently been claimed that natural selection, which can explain so much, is vacuous, but in the words of Beckner (1959):

> No discredit is cast upon selection theory by showing that it is in fact compatible with all available evidence. On the contrary, discredit would accrue only if it were shown to be compatible with all possible evidence.

As Reed (1981) has explained further, the universal explanatory power of natural selection should not be taken to mean that it is universally applicable.

If we wish for a truly vacuous explanation of the state of the natural world, there is Philip Gosse's (1858). This was described by his son Edmund Gosse (1907) as follows: the elder Gosse believed that there had been no gradual modification of the surface of the earth, or slow development of organic forms, but that when the catastrophic act of creation took place, the world presented, instantly, the structural appearance of a planet on which life had long existed.

This explanation will not be discussed further in this book. (As Gibbon (1776) put it:

There are some objections against the authority of Moses and the wisdom of the prophets which too readily present themselves to the sceptical mind, though they can only be derived from our ignorance of remote antiquity and from our incapacity to form an adequate judgement of the divine economy . . .)

The general plan which will be followed is first, to define natural selection and some other hypothesized broad mechanisms of evolution; secondly, to discuss general constraints on any evolutionary process; thirdly, to discuss hypotheses about very early evolution, i.e. soon after the origin of life (but not to treat that origin, except by implication); and finally to consider evolution as constrained by something rather like the genetical mechanisms existing today. This last topic constitutes the bulk of this book, since it is here that natural selection may be assessed most satisfactorily as a mechanism of evolution.

1.1.1 natural selection

Natural selection means the differential viability and fertility of different members of a population as a result of their different degrees of adaptation to the environment. The name is an analogy. As Nagel (1979) wrote:

Natural selection is not literally an "agent" that *does* anything. It is a complicated process, in which organisms possessing one assortment of genetic materials may contribute more, in their current *environment*, to the gene-pool of its species with different genotypes.
In short, natural selection is "selection" in a Pickwickian sense of the word.

Actually, it is in a Darwinian sense: Darwin chose the analogy, and as he wrote of his work on evolution in the late 1830s,

I soon perceived that selection was the keystone of man's success in making useful races of animals and plants. But how selection could be applied to organisms living in a state of nature remained for some time a mystery to me.
(Darwin, 1882)

Natural selection was also an analogy which was in the air; apart from Chambers and Erasmus Darwin, others were thinking in similar vein:

For Nature also, cold and warm,
And moist and dry, devising long,

> Through many agents making strong,
> Matures the individual form.

as Tennyson wrote in 1833.

The argument that natural selection is tautological because it can be compressed into Herbert Spencer's phrase "the survival of the fittest" has occupied too much of people's time (cf. Popper, 1957, 1974; Ruse, 1977), so I shall mention only Lewontin's (1969) counter to this argument. He pointed out that evolution was a necessary consequence of three verifiable observations. First, variation occurs. Secondly, offspring resemble their parents. Thirdly, "different phenotypes leave different numbers of offspring in *remote* generations". Add to this the further observed fact that some of the variation is advantageous, some disadvantageous and this is the essence of Darwinism.

Problems arise in applying the simple principles inherent in this argument. Wasserman (1978) has argued that there are two uses of natural selection in evolutionary explanations. First, it may be included in an axiomatically based population genetics. Secondly, it may be used *ad hoc* to "explain" particular adaptations. This second usage is less than respectable (cf. Lewontin, 1974, and Chapter 5). However, Wasserman also states that

> in the few cases where selective factors have been found these have been obtained by methods which are specific in each case. I conclude that the discovery of particular selective factors by a method specific for each factor is no more useful to a theoretician than the discovery of those isolated cases of natural selection which have no bearing on a population genetic theory.

It is not clear why this should be so; why should the factors leading to heavy metal tolerance in *Agrostis tenuis* (Bradshaw *et al.*, 1965) be the same as those causing great fluctuations in the frequency of a gene affecting wing-spotting pattern in *Panaxia dominula* (Fisher and Ford, 1947; Ford, 1975)?

Natural selection is a very general term describing a vast range of possible mechanisms. No concealed *ad hoc* assumptions should be made when the determination of one specific phenomenon is achieved, but the factors involved in a particular case will certainly be specific *ad hoc*.

The hypothesis of evolution by natural selection which is advanced in this book is that natural selection of individual differences, large or small, is the main driving force of evolution. Constraints on natural selection, and variants of natural selection which relate more to groups of related individuals than to single individuals, are the main topic.

We need, however, to consider the other mechanisms which have been seriously advanced, noting that many evolutionary hypotheses, for example about the origin of a major taxon, are not hypotheses about mechanism at all. Van Valen and Maiorana (1980) point out that such hypotheses may be tested in a sense by drawing up a topologically satisfactory phylogeny, and continually stressing it by including more traits. But the events leading to the occurrence of four different cell-wall types in archaeobacteria (Fox *et al.*, 1980) compared with but one in the far more numerous eubacteria, may never be reconstructed in such a way that the role of mechanisms such as selection is elucidated.

1.1.2 orthogenesis

Orthogenesis is the process, associated with the names of Bergson (1907) and Osborn (1909, 1921, 1922), whereby evolutionary change appears to be towards some goal, as if endowed with what in human affairs is termed purpose. Gurwitsch (1915) suggested that something analogous to an electro-magnetic field might yield the appearance of purpose, but, like other writers in this area, provided no evidence for such an "immaterial" factor. Indeed, no mechanism has yet been postulated for such a process, but Fisher (1929) was able to suggest how selection for the genetical phenomenon of domi-nance might accelerate once it had begun, giving the appearance of movement towards a goal. Dominance and recessiveness of traits had been introduced as concepts by Mendel (1865) to explain the occurrence of two distinct forms of a trait in the offspring of an individual of one particular form. These concepts were taken for granted by many geneticists, i.e. regarded as inherent properties of genes as displayed in the phenotype, but Fisher in 1928 had suggested that dominance might have evolved through the accumu-lation of genes modifying the heterozygote to resemble one homozygote. In the ensuing controversy with Wright (1929; see Sved and Mayo, 1970, for discussion), Fisher showed that the process of modifier accumulation would accelerate, at least for a time, as the frequency of modifying genes rose. (These processes are discussed further in Section 7.3.4.) At the time, this was a novel idea, but it is now taken for granted, as in the results in Table 6.1 (see Chapter 6) for standard, well-understood gene frequency change; such cases were of course not known to Bergson and Osborn.

This is, however, not the only possible directed change. Others include those processes of accumulation of DNA probably inherent in the meiotic process and perhaps meiotic drive. The latter is a mechanism affecting the genetical composition of a population through inequality in chromosomal transmission during meiosis, so that genes on a preferentially segregating chromosome make a disproportionate contribution to the next generation. It has yet to be demonstrated conclusively to have been important in micro-evolution.

1.1.3 pre-adaptation

Associated with the concept of orthogenesis is that of pre-adaptation (Cuénot, 1951). It may be summarized in Davenport's words (Cuénot, 1914, cited by Fisher and Stock, 1915):

> The structure exists first, and the species seeks to find the surroundings which respond to its particular constitution. The adaptive result is not due to a selection of structures suitable to a given environment (theory of Darwin and Wallace), but on the contrary to the choice of surroundings responding to a given structure.

As Fisher and Stock pointed out, "For 'species' read 'individual', and a grain of truth emerges." In this sense, one might simply say that mutant individuals better adapted than others to a particular small part of their environment and able to move to that part of the environment may do so and thereby gain an advantage. A species, however, cannot change its environment in the same way.

Darwin (1868) had recognized that distinct mutations occur, as when he wrote about poultry: "The tarsi are often feathered. The feet in many breeds are furnished with additional toes. Golden spangled Polish fowls are said to have the skin between the toes well developed." However, he did not consider that the mutations themselves yielded new varieties and thence new species. De Vries (1902; cf. Davenport, 1909) suggested that new species might arise through large mutations, but, unfortunately, his evidence came from a faulty understanding of segregation patterns in *Oenothera lamarckiana* (Jeffrey, 1915). In Lamarck's primrose, segregation as a result of the unusual genetical system regularly gives a small proportion of very distinct, discrete character changes, which breed true on self-fertilization. In essence, however, these are no different from the modal types of the species than are any ordinary homozygous recessives in any other species; they are certainly not "elementary species", as de Vries called them. Evidently, a confusion had arisen between the origin of heritable variation and the process of speciation.

Following these ideas, however, Goldschmidt (1940) and others suggested that new forms might arise differing substantially from the mode, but highly adapted to a new way of life. Mimicry was chosen as an example; a mimetic form, resembling some model in appearance to a remarkable degree, and adapted behaviourally to mimic that model, would arise by a saltation, or leap, and thereby create a previously non-existent niche. However, mimetic patterns have a complex genetical origin. Rather than being determined by allelomorphic genes, as might have been supposed superficially, they prove on detailed analysis to be determined by many loci (at least six in the butterfly *Papilio memnon*), usually tightly linked as a supergene or gene complex (Clarke and Sheppard, 1963, 1971, 1973, 1977; Clarke *et al.*, 1968). The mimetic pattern is improved further by other loci in the gene complex unassociated by linkage with such supergenes (Sheppard, 1969). Canalization may also make the joint contributions of many genes appear to be a macro-mutation (see Section 7.3).

Gans (1979) noted that adaptation will be achieved with a range of phenotypes, since environments are rarely constant. Some of these phenotypes will be what Gans called excessive in production of whatever is required for adaptation, and some will be inadequate. If then migration or environmental change allows a population to enter a new adaptive zone (Simpson, 1953), directional selection on the trait of interest, i.e. an advantage occurring to individuals in proportion to the magnitude of this trait, may result in its stabilization at a level which was previously excessive. Thus, it will appear that the part of the phenotypic range previously excessive was "pre-adapted" to the new environment. Gans has called this "proto-adaptation", to distinguish it from pre-adaptation in Cuénot's sense; it is not a novel process.

1.1.4 evolution of acquired characters

This is a process which occurs, if it does, as a result of natural selection, but more directly than in the widely accepted view of selection acting on random mutation. As originally proposed by Lamarck (see Jacob, 1970), organisms changed, to some extent purposefully, in response to environmental stimuli, and these changes were then inherited. Although results are obtained from time to time which appear to suggest that such a set of events has occurred, they have rarely been repeated satisfactorily (Gorczynski and Steele, 1980, 1981; Brent *et al.*, 1981; McLaren *et al.*, 1981; Smith, 1981; Durrant, 1958; Cullis, 1977). The regularity of meiosis and the difficulty of somatic changes altering germ-line DNA seem to provide effective barriers to frequent Larmarckian evolution in present-day organisms.

However, before the one-way pattern of information transfer evolved to its present precision, environmental modification of the genetic material could have been a significant process.

1.2 THE CORPUS CALLOSUM AS A SPECIFIC EXAMPLE OF THE PROBLEM OF EVOLUTIONARY EXPLANATION

The corpus callosum have been defined since early last century as the midline commissure of the cerebral hemispheres (Joynt, 1974). Its function has long been the subject of debate. As Wood Jones (Wood Jones and Porteus, 1928) put it:

> The question of the corpus callosum is historic. Battles were waged concerning it; battles instructive of the times in which they were waged, and in which might has not always been with the right. The days of the fight are done with and we say that the development of a true corpus callosum takes place only in the higher, non-marsupial, mammals. The marsupials have as their hallmark this failure, that the corpus callosum – the great, dominant, fibre tract connecting the all-important non-olfactory parts of the pallium did not develop in their phylum. Here then is [a] story of failure in brain building. The lowest mammals, represented by the monotremes, possessed a large ventral commissure and a relatively small dorsal commissure. The marsupial developed the dorsal commissure to a greater degree but failed altogether to effect the structural modification which begets a true corpus callosum. They are failures. Wherever marsupial meets higher mammal it is the marsupial that is circumvented by superior cunning and forced to retreat or to succumb.
>
> Since the corpus callosum is obviously of such great importance in the phylogenetics series of mammals, it might be expected that there would be abundant experimental evidence as to the exact role that it played in bringing about the advance of those mammals in which development was initiated.

Knowledge, however, was limited. The precise interconnections mediated by the corpus callosum had been demonstrated; acallosal but apparently normal persons had been identified at autopsy; split-brained animals were unable to integrate perceptions and conditioned reflexes. But this was hardly sufficient evidence for Wood Jones's conclusion that

> complete integration of the whole individuality in cortical terms is the triumph of the human brain. This triumph was made possible by the development of the corpus callosum from the dorsal commissure in the lowly monodelphian mammals.

Leaving aside the question of the "olfactory" portions of the pallium (Smythies, 1970), and the fact that marsupial brains do not differ dramatically in most other aspects of gross morphology from eutherian brains (Meyer, 1981), we see here a remarkable argument from correlation to causation where, because of Wood Jones's remarkable insight, his logically invalid inference – that because the corpus callosum is associated with the evolution of higher cerebral functions it must be necessary for those functions – is now seen to be correct (Table 1.1), on the basis of a great deal of experimental evidence not relevant in detail here. (The table could be extended in a number of ways, for example, to include interhemispheric transfer and perhaps memory.)

If Wood Jones's phylogeny were accepted, he had demonstrated evolutionary development of the corpus callosum; he had not demonstrated selective advantage, except indirectly, anecdotally, by analogy. Further-

Table 1.1   The function of the corpus callosum

Author	Date	Function
Galen	*ca.* 200	none attributed
Vesalius	1543	structural support
Willis	1664	seat of the imagination
Vicq d'Azyr	1784	"sympathetic communications between different parts of the brain"
Gall	1810	specific interconnections of equivalent regions of the two hemispheres
Dejerine	1892	transmission of specific information between specialized regions
Wood Jones	1928	"correlating the two sides of the body"; integration of perception
Abbie	1941	"essential to the most effective functioning of the forebrain on a whole"; necessary for "the progressive cerebral dominance which is the main feature of eutherian evolution"
McCulloch	1944	"a mystery"
Lashley	1950	relatively minor in "the associative connexions or memory traces of the conditioned reflex"
Bremen	1956	"highest and most elaborate activities of the brain"

Summarized from Joynt (1974), Wood Jones and Porteus (1928), Abbie (1939, 1941).

more, only the relatively gross defects were known to him, and some of these, as noted in Table 1.1, appeared to have relatively minor effect. More precise investigations, however, have shown that minor damage to the corpus callosum produces minor but detectable deficiency in normal perceptual or cognitive processes (cf. e.g. Kinsbourne and Smith, 1974). These deficiencies have not yet been shown to be critical for reproductive success and hence Darwinian fitness. The chain of explication is incomplete.

Is any other model more powerful in explanatory theory? Is a large link between the cerebral hemispheres necessary for development but less important in the adult animal? Evidence suggests not (Kinsbourne and Smith, 1974). It is not clear whether our inability to derive useful alternative hypotheses is a result of poverty of imagination (contrary to the suggestions of Brady and Lewontin cited earlier) or to an inability to think past the Darwinian model.

TOPICS FOR DISCUSSION

1. "The tenacity of the wonderful fallacy that the laws of nature are agents, instead of being, as they really are, a mere record of experience, upon which we back our interpretations of that which does happen, and our anticipation of that which will happen, is an interesting psychological fact; and would be unintelligible if the tendency of the human mind towards realism were less strong." (Huxley, 1887a).

    What bearing does this idea have on our interpretation of the concept of natural selection?

2. "The key of the past, as of the future, is to be sought in the present; and, only when known changes have been shown to be insufficient, have we any right to have recourse to unknown causes." (Huxley, 1887*b*).

    Does the concept of orthogenesis conflict with this precept?

3. Fluoroacetate poisons have been developed to destroy certain small mammals, such as rabbits, in Australia. Native mammals, both eutherian (e.g. *Rattus fuscipes*) and marsupial (e.g. *Macropus eugenii*), have been found to be tolerant to these poisons (Oliver *et al.*, 1979). The legume genera *Gastrolobium* and *Oxylobium* naturally contain fluoroacetate, and form part of the diet of the relevant animals.

    Is fluoroacetate tolerance a pre-adaptation?

2 general boundaries for evolution

Some aspects of our universe, the only one available to us for investigation, impinge on possible paths for evolution, no matter what the form the evolution has taken. In this chapter, I examine three important aspects of the real world for their relevance to the form and extent of natural selection.

2.1 TIME

Evolution has occurred; the evidence is too strong to believe otherwise. This is a truism, but despite it, there are questions of some importance about the time which has been available for evolution. The problem has been stated simply by Salisbury (1969):

> a typical small protein might contain about 300 amino acids, and its controlling gene about 1000 nucleotides (3 for each amino acid). Because each nucleotide in a chain represents one of four possibilities, the number of different kinds of chains is equal to the number 4 to the power of the number of links in the chain; that is, 4^{1000}, or about 10^{600}.

For every one of these amino acids to be chosen by natural selection to be optimal, or near optimal, in only 4000 million years requires a quite extra-ordinary intensity of selection, while the possibility of an enzymatically active, stable molecule arising by chance is so close to zero that it can be treated as such. Salisbury suggested that there are two ways out of the problem. One possibility would be that only a few of the amino acids in a protein were essential for its function. There is considerable evidence, certainly, that not all positions in a polypeptide chain are equally important. However, given that there are at least several thousand different poly-peptides in any advanced eukaryote, the problem remains.

Salisbury added that "Special creation or directed evolution would solve the problem of the complexity of the gene, but such an idea has little scientific value in the sense of suggesting experiments." However, Maynard Smith (1970) proposed a possible solution which could be regarded as a kind of directed evolution, or at least a channelled or constrained evolution. He suggested that if all existing proteins are part of a continuous network, i.e. that all functional proteins form a space in which movement from one functional protein to another is a matter of a small number of finite mutational steps, then given the initial extraordinary chance of a functional protein arising, the development of many others therefrom will not be as improbable as it now seems. The initial production of a functional protein, in fact, may not itself have been such an extraordinary event if protein evolution was initially a process of accretion, small polypeptides of very limited activity gradually becoming incorporated into larger molecules with increased stability and other advantageous properties, partly through duplication of the genetical material. (From the minimal values shown in Table 6.4 (Chapter 6) for change within proteins, there has certainly been time for every amino acid to have been replaced by another at least once, purely by chance, in most proteins, and changes to the information-bearing nucleic acid have been even more rapid, whether as a result of selection or drift.) Evidence for these views lies in the fact that the three-dimensional structure of proteins, having a given function, has been strongly conserved, even when the basic polypeptides are very different, thus illustrating how divergence can occur within the "protein space" (cf. e.g. Matthews et al., 1981). However, it appears that changes in structure have not necessarily (nor even usually) been along the shortest possible path (Holmquist, 1972; Holmquist and Cimino, 1980).

It will be recalled that Darwin's evolutionary time-scale was such that he considered evolution must have taken several hundred million years at least, and that this was contradicted by the work of the nineteenth century physicists, especially Lord Kelvin. In fact, it was the physicists who were at fault rather than the biologists in that particular case; they had not taken into account, and of course can hardly be criticized for not taking into account, the residual radioactivity in the rocks of the earth and its contribution to the temperature of the earth. However, our current knowledge of the evolution of life suggests that it is a process which began certainly more than 2300 million years (Cloud and Morrison, 1979), probably more than 3400 million years but less than 4000 million years ago (Lowe, 1980; Walter et al., 1980).

Given this amount of time, one is almost forced to postulate some form of very rapid, pre-biotic and early biotic evolution, since the earliest recognizable life-forms have, obviously, many of the characteristics of later fossil, or even current, life-forms. (If they did not, we would hardly recognize them.) Thus, the first 600 million years of evolution must have contained a period in which present biochemistry evolved. In fact, the process may have been very rapid indeed (see Chapter 4). On the other hand, it is not clear that the rate was rapid during the succeeding 2500 million years, since Metazoa appear in the fossil record no more than 700 million years ago (cf. Cloud and Morrison, 1979; Cloud et al., 1980).

In general, given our ignorance of early evolution, we can only conclude that evolution through natural selection, as an organizing process, can account for what appears to have occurred in the apparent time available, provided that early evolution both occurred very rapidly and yielded a reasonably effective information storage and transmission mechanism. The former requirement appears to be accommodated by the hypotheses of early evolution discussed in Chapter 4, and the latter is reasonable given the near-universality of the genetic code and its translation and transmission mechanisms. The diversity of extinct and present life-forms also points to the rapidity with which change can occur, although the bearing of the existence of such diversity on the role of natural selection is more related to specific questions of selective efficacy in already highly developed organisms than to early evolution.

2.2 INFORMATION

It has been suggested that "the total informational content of the biosphere [is] a quantitative measure of functional efficiency in processing information and . . . an objective criterion of evolutionary progress" (Theodoridis and Stark, 1969). Further, it has been suggested that "the structure and the function of living organisms can be best described in terms of the physical quantity of information" (Theodoridis and Stark, 1971). I shall show that the quantity of information may be a misleading parameter to consider in genetical evolution, since it takes no account of different kinds of information, nor does it possess any kind of predictive value, which is perhaps what is also missing from the theory of evolution by natural selection, which measures everything *a posteriori*.

Theodoridis and Stark (1969) present the following scheme: let us consider that solar flux is the main source of information (in the form of energy) for the evolving earth. Then a quasi-steady-state situation exists and the solar information input i_s can be divided into a biospheric part i_b and a non-biospheric part i_{nb}. Now if τ_b and τ_{nb} are the biospheric and non-biospheric information life-times, the biospheric information I_b is given by $\tau_b \simeq I_b/i_b$, and similarly $\tau_{nb} \simeq I_{nb}/i_{nb}$. Then $I_b \simeq \tau_b(i_s - i_{nb})$. It is clear immediately that as we do not have estimates of τ_b and i_{nb}, we cannot estimate I_b; furthermore, as Theodoridis and Stark (1969) note, cataclysmic events such as earthquakes, etc. affect I_b without materially affecting i_s, so temporal comparisons may well be meaningless.

To see how informational arguments may be misleading in genetic evolution consider the simple case of two competing species whose information contents are I_1 and I_2. If species 1 replaces species 2 (and species 2 becomes extinct), this does not imply that $I_1 > I_2$, except in a trivial sense; in any case, we do not know how to measure I for a given species. (The amount of DNA could be used as a criterion (Kimura, 1961), but there is evidence that closely related species can differ substantially in this measure (Hayman and Martin, 1969), and furthermore not all DNA is equivalent.) When in about 1680 the last dodos died without issue, their eggs eaten by pigs or dogs (Strickland and Melville, 1848), some information was lost; unless a temporary part of the increase in abundance of these predators inadvertently introduced, for other purposes, by man is counted, it is difficult to argue objectively that there was a biospheric information gain in this disappearance, even though this highly specialized and not numerous species was probably unlikely to survive long. While less than 1% of all species ever arising are still extant (Mayr, 1963), the loss of many species (not just the dodo) represents a loss in information not necessarily regained as suggested (Theodoridis and Stark, 1969). For example, during the last ice-age, many species of trees and several of the great mammals were extinguished (Zeuner, 1946), and there is no real evidence to suggest that they or their descendants would all, inevitably, have proved inadequate competitors, as (most probably; cf. p. 7) were the South American marsupial carnivores when North and South America became joined in the upper Pliocene (Mayr, 1963). (Extinction will be considered in more detail in Chapter 10.)

Evolution by natural selection acting on genetically determined variation acts to increase information within a species (Fisher, 1930), but does not necessarily do the same between species, since it has as characteristics both adaptive radiation (increasing variance) and frequent extinction (decreasing variance). The process of acquisition of information in genetical (as opposed to pre-genetical (Theodoridis and Stark, 1971)) evolution is far removed from the steady state information gain by solar flux. Rates of evolution in particular groups of animals have by no means been constant; for example, the lungfishes changed as much in the 30 million years of the middle and late Devonian as they have in the 250 million years since (Mayr, 1963). With any particular species, the rate of acquisition of information may well decline as the amount of information increases (Lewontin and Waddington, 1967), though this need not always be the case (Mayo, 1971).

It has been suggested that "the viability of a new biological species . . . can be assessed by determining whether [its] survival results in a biosphere of a higher information content" (Theodoridis and Stark, 1969), but apart from problems of measurement this is to reverse the problem or reduce it to the tautology dear to the critics of the theory of natural selection, since survival of a species is the measure of viability. Furthermore, as already indicated, survival of one species, if at the expense of another, need not result in increased biospheric information content.

If the biospheric information content were really an objective criterion of evolutionary progress then it should be possible to derive minimum conditions for the existence of life (however defined) and hence to establish the time at which life began, assuming it not to have an extra-terrestrial origin. However, we do not yet know how much information there is in the bio-sphere, nor what solar input is necessary to maintain some life (apart from human use of stored energy); perhaps the current solar flux is the least upper bound of information input for the maintenance of the biosphere in its present state. Even this is not certain; as Haldane (1937) recognized many years ago, in macro-evolutionary terms it is necessary to consider such factors as the possible decline in **g**, the gravitational constant, by as much as one part in 10^9 per year (Dirac, 1937; Kuchowicz, 1971), though this par-ticular factor has probably been without significance. At the moment, it is not clear whether more or less genetic information is necessary to cope with a reduction in the strength of the gravitational field. In any case, the evolved conservatism of genetic systems (Darlington, 1939; cf. Sections 7.3 and 6.5) can be expected to retain obsolete information for long periods of time.

Overall, it can be stated that the solar flux of information cannot yet provide any criterion for genetical evolution, and possibly not for pre-genetical evolution either (McClendon, 1980). Results yielded by the information–theoretical approach (Theodoridis and Stark, 1969, 1971) suggest essentially that we do not have enough data to apply the method properly. More precise definitions of varieties of biological information appear necessary. Its use at the level of the genetic code, however, can be helpful (cf. Section 6.3).

2.3 COMPLEXITY AND ORDER

In the previous section, it was shown that information flow is not an ade-quate or sufficient measure of evolutionary change. This inadequacy has to some extent been recognized (cf. Papentin, 1980, for discussion). Con-sequently, further concepts have been introduced which might be regarded as having more relevance to biological evolution. One of these is com-plexity.

Complexity may be considered at two levels. One is the organism: does evolution necessarily bring about increasing complexity (however defined)? Secondly, is increasing complexity at the eco-system level a stabilizing phenomenon, and how does this affect evolution by natural selection?

Papentin (1980) defines complexity as "a special way of estimating the information content of a pattern", i.e. the additional concept of pattern is necessary. Thus, complexity will consist of two parts, organized complexity, relating to the "rules or laws underlying the formation of a pattern" and unorganized, relating to random aspects. Then a definition for analytical purposes of complexity is given by Papentin as the minimal description of a pattern. This will in general not be available, since we will not know what to include. Papentin confronts this problem but does not solve it.

Saunders and Ho (1976, 1981) define complexity differently, first as the number of different components in a system and later as a variant of the information statistic considered in the previous section. However, they also suggest that it is entirely relative, in fact observer-determined. Accordingly, it is difficult to accept their suggestion that Williston's rule may be deduced from the ideas that complexity, not fitness, is the increasing function which gives direction to evolution and that the observation of a particular feature in evolution "largely depends therefore not on its relative selective advantage but on the probability that it occurs in the first place." This is evolution driven by mutation in a different guise; Saunders and Ho, however, make the point that small changes are more probable than big ones, and that reiterated structures have more opportunity for change than singly represented structures. Hence, change, either modification or decline, is to be expected; and this is Williston's rule. Their argument, however, is really a judgement that random change is more likely than selected change, and should be based on the evidence available from population genetics (cf. Chapter 6), rather than on a theoretical model of mutation frequency.

We turn now to the complexity of eco-systems.

Felsenstein (1978a) has considered a very simple model eco-system which evolves a balance of energy flows through a number of trophic levels, whereby part of the loss of energy from each level consists of a flow to the next level (with flow unidirectional by the second law of thermodynamics). Then the more levels in the system, the more energy is retained in the system. The more unpredictable the environment, the less energy is retained in the system. The first of these conclusions implies that complex eco-systems have an advantage, if increasing information (i.e. energy content) is advantageous. The second appears to imply that evolution will be slowed by an unpredictable environment.

Rutledge et al. (1976) have considered a slightly more specific model of ecological stability (defined as "the ability of an ecosystem to resist changes in the presence of perturbations") which appears to imply that stability is dependent on diversity of "metabolic structure", but this might be observed within an organism (or species) not just a whole eco-system. Tregonning and Roberts (1979; Roberts and Tregonning, 1980) have examined a collection of m species interacting together and varying in number N_i on the generalized Lotka–Volterra model

$$\frac{dN_i}{dt} = N_i (r_i + \Sigma A_{ij} N_j), \quad i = 1, \ldots m,$$

where the r_i are intrinsic rates of increase and the A_{ij} are interactions between species i and j. For this simple model, they have shown that for small m the equilibrial values for A_{ij} and r_i correspond to reasonable real cases, e.g. predator/prey interactions with $A_{ij} > 0$, $A_{ji} < 0$, $r_i > 0$ and $r_j < 0$. They have further shown that natural selection, i.e. extinction of some species through selection, is a more rapid path to equilibrium than random extinction. A very remarkable result of their analysis (and one needing further investigation) is that a robust eco-system is one consisting of sub-systems which are minimally inter-dependent. Experimental results or observations relating to this last result are extremely rare, and are not much more frequent for the general question of the relationship between diversity and stability, though such relations have long been sought (see Woodwell and Smith, 1969). Indeed it has even been argued that species communities (which are not whole eco-systems, of course) may be assembled largely at random (Connor and Simberloff, 1979).

Overall, from these results, one might perhaps conclude that progress under natural selection may be most rapid in a relatively constant environment, a result not much accepted at the present time, and that stability of an eco-system will be the result of natural selection. Evidently, such general arguments have little force as yet.

TOPICS FOR DISCUSSION

1. What assumptions must be made in order to quantify Salisbury's hypothesis, of insufficient time for evolution by natural selection to have produced the present state of nature, to make it testable?

2. What criteria must be fulfilled in the construction of a measure of complexity in an organism to make that measure useful for the comparison of the evolutionary development of different organisms?

3. "A eukaryotic cell can be likened to a society composed of a nucleus and a crowd of subcellular organelles in which all members cooperate for the common good." (Eberhard, 1980).

   Under these circumstances, what would one predict for the organization of the genetical determination of structure and function by nucleus and organelles? Is the presence of separate DNA in nucleus and organelle predictable? If not, what additional hypothesis is needed to make this fact predictable?

3 physical constraints

Apart from divine intervention, all mechanisms postulated for the state of nature which we observe are bound by the same physical constraints. Nevertheless, examination of those constraints may show first, where they are specifically relevant to natural selection and secondly, where other mechanisms may not be restricted.

3.1 OVERALL CONSTRAINTS

Most of these constraints relate to simple physical quantities. For example, generation interval and life span are related both to the effluxion of time and to the size of organisms. The question of life span is discussed in Section 8.3. The mass and hence the size of an organism are evidently limited by gravity. This was first discussed in some detail by Haldane (1928). More recently, Economos (1981) has suggested that there is a gravity-determined upper limit of approximately 20 000 kg for a land mammal, the approximate size of the largest known mammal, the extinct Baluchitherium. (Different rules evidently applied to Dinosaurs, since *Brachiosaurus*, for example, was about four times as large, but certainly not because of the variation in **g** mentioned in Section 2.2.) A lower limit of 0·0025 kg was suggested by Pearson (1948), on the basis of metabolic requirements; and some shrews approach this limit. Thus, size, as an aspect of the phenotype space, is fully explored, but where the boundaries exist, they exist regardless of the agent of evolutionary change. There are other constraints on size; e.g. the overall size of insects is limited by the diffusion properties of oxygen and carbon dioxide, as a result of their mode of respiration.

The velocity at which animals can travel is limited by the strength of the materials of which they are constructed and by the power of their muscles. So are other aspects of mobility and activity. The constraints on all these traits are not simple. For example, it is well known (cf. e.g. Maynard Smith, 1968) that the power output of animals is approximately proportional to the square of their overall length, whereas mass varies as its cube, so that different types of animals move, proportionately, very differently, e.g. a flea leaps many more times its height than does a kangaroo. None the less, the evolutionary restrictions appear only to be the boundary type for size discussed above.

Plants and animals have evolved essentially different means of coping with variation in temperature, but have the common constraint that normal metabolism can only occur within a limited range of temperatures. For animals, this imposes the additional constraint that for occupation of niches at different temperatures, different body sizes are necessary (see e.g. Searcy, 1980). The required internal operating temperature of an animal has also been claimed to be far higher than would be optimal for information processing, thereby, perhaps, imposing limits of the efficiency of sensory and cognitive processes (Ball, 1980).

An overall physical constraint of a different kind, one related to the nature of genetical mechanisms as these have evolved, is that of sex. It was pointed out by Fisher (1930*b*) that a species composed of more than two sexes must be genetically unstable; organisms are restricted to one or two sexes. Furthermore, the existence of sex imposes the constraint that advantage to an organism, if it is to persist, must be of such a kind that some of it at least will persist through the randomizing process of reduction division. This problem is discussed further in Chapter 9.

3.2 ALLOMETRY

Allometry has long appeared to be an important constraint on evolutionary change, i.e. as Thompson (1917) and Huxley (1932) first pointed out, the size of an organism to some extent determines its shape and function, both in whole and in individual organs. Huxley (1932) concluded that a power function

$$y = ax^b.$$

was a law of relative growth, the parameters a and b describing the form of the relationship in any given case.

This was soon recognized to be an oversimplification (Thompson, 1942; Yates, 1950; Smith, 1980). However, in some cases it may reflect underlying causal mechanisms. For example, Apple and Korostyshevskiy (1980) have suggested that basal metabolisms are related by a power law since basal metabolism varies linearly with the number of mitochondria, and body weight varies linearly with the number of cells, and cell and mitochondrion number are regulated co-ordinately.

Investigations of the genetics of allometric relationships were begun very early, especially for eugenic purposes, but were often vitiated by severe problems in data collection and analysis (cf. Fisher, 1936; Fisher and Gray,

1937). More recently, many studies have shown that allometric relationships may be highly heritable with substantial present genetical variability (Atchley and Rutledge, 1980).

When a particular pattern of dependence exists, e.g. that between brain size and bodyweight in primates (Clutton-Brock and Harvey, 1980; Harvey *et al.*, 1980), departures from the relationship may be related to major inter-family differences, even though the form of association between brain size and adaptation has not yet been elucidated. This is similar to the case of the corpus callosum discussed earlier, but potentially testable hypothetical mechanisms have been advanced. An example is the suggestion of Clutton-Brock and Harvey (1979) that animals which exploit unevenly and widely dispersed but clustered food supplies will require larger brains than others subsisting on more predictable food sources. This has not yet been proved, and the problem of testing a single adaptive hypothesis is very complex.

Nevertheless, we see that the constraint of an allometric relationship can be readily modified in principal by natural selection. Modest initial changes in allometry may indeed relate to major evolutionary departures, as Fisher (1954) suggested. If an "ontogenetic trajectory" (Alberch *et al.*, 1979), i.e. a developmental pathway, is altered, the outcome may not be substantially different initially, but the potential for change is extremely different. The evolution of the wing from the forelimb, a process which has occurred independently three times, is a dramatic example. The evidence for the importance of trajectory changes is limited as yet (see e.g. Hampé, 1959), but as a source of variation on which selection might act, allometric variation is a process of great potential. Other models of evolution appear to rest less easily with the facts, in that regular processes might be expected to be least susceptible to modification by major saltatory changes.

3.3 LIFE HISTORY

An organism in a particular ecological niche has considerable constraints on its potential direction of change on account of its life history.

Restraints on size are not limited only to those imposed by the physical nature of the world. For example, viviparous animals are restricted in birth size by the size of their parents and this may well influence their final maximum size.

Similarly, life span may depend on life history. Plants which are long-lived perennials cannot also be ephemeral annuals, thereby colonizing a different niche. Annuals may also be perennial, but only to a limited extent.

Time to maturity, such as the extremely prolonged development time of the human infant, depends upon the life-time of parents. The two are likely to increase in step.

Fecundity is also limited by the proportion of resources which parents may contribute to the production of offspring. The major "strategies" found in different organisms are r-selection and K-selection (MacArthur and Wilson, 1963, 1967). In the former, the number of offspring is maximized at the expense of parental care, in the latter, the reverse is true. Plants are evidently mostly at the r-end of the spectrum between the two types, while animals of all possible types exist. Indeed, almost any pattern which can be hypothesized can be shown to have an advantage under certain conditions. For example, Heckel and Roughgarden (1980) presented a model of r and K interaction such that a population regulated in density by the environment could undergo change towards a lower intrinsic rate of increase (r), despite the general tendency to increase the efficient use of resources and hence increase population density. The reasons for these results are not obvious biologically. This example illustrates the problem of considering life history outside a framework of different types of selection.

Many life-history constraints appear to be related to stabilizing or normalizing selection (cf. Chapter 7). For example, Roff (1981) has demonstrated in some members of the genus *Drosophila* a strong, curvilinear relationship between body size and the rate of population increase (r), with an optimum size, in terms of maximizing r, near the mean of the size distribution. However, the extremely large Hawaiian *Drosophila* do not lie in the same range, and their size therefore cannot be accounted for in terms of the same relationship between size and reproductive rate.

3.4 DIRECTIONAL CHANGE

In Chapter 6, rates of change under natural selection and random genetical drift will be considered. The question of interest here is to what extent small changes may be of no physical benefit, and large changes may not be practicable.

The largest known bird, now extinct, is thought to have had a wing span of as much as seven metres. The largest known flying creature, on the other hand, *Quetzlcoatl northropi*, a pterosaur, is thought to have had a wing span of as much as eleven metres (Lawson, 1975; Stein, 1975; McMaster, 1976). It seems that the pterosaurs had a different mode of flight, involving a very low stalling speed, but with take-off only possible in light but steady winds (Hanbin and Watson, 1914; Bramwell and Whitfield, 1974). The range of types of bird flight does not include this category. Would it therefore have been possible for a bird to have evolved by small increments from the range of wing-loadings known (Greenewalt, 1975) to a pterosaur with its much lower wing-loading? This appears unlikely; is this a rare "anti-allometric" case where a mutation increasing overall size by a very substantial margin might have allowed a bird similar to a pterosaur to have evolved? This does not appear to be how the pterosaurs themselves evolved (cf. e.g. Lawson, 1975).

Consider, on the other hand, the question of a small increment in wing area. I shall follow the analysis of Greenewalt (1975) (cf. also Pennycuick, 1972).

By a process of trial and error, Greenewalt divided birds into three groups on the basis of wing loading for a given weight. The models are named after the most important groups which they contain, viz. the "passeriform" model, which includes herons, falcons, hawks, eagles and owls, the "shore bird" model, which includes doves, parrots, geese and swans as well as "shore birds", and the "duck" model, which includes grebes, loons and coots. The relationship between weight, W, and wing area, S, is as follows:

Passeriform	$W \doteq aS^{1 \cdot 274}$	$0 \cdot 032 \le a \le 0 \cdot 080$
Shore bird	$W \doteq 0 \cdot 044S^{1 \cdot 4}$	
Duck	$W \doteq 0 \cdot 082S^{1 \cdot 4}$	
Bat	$W \doteq 0 \cdot 014S^{1 \cdot 524}$	

Considering the relationship between wing loading and total weight, Greenewalt compared a very large number of birds, also bats and some insects, with the Boeing Company's 700 series jet aircraft, concluding "Apparently nature follows the rules for dimensional similarity in an overall sense, but reserves (and indeed exercises) the privilege of creating models which depart substantially from these requirements for portions of the overall size range." For example, "dimensional similarity would require a constant aspect ratio; we see that for birds the aspect ratio increases with increasing size." (Aspect ratio is defined as (wing span)^2/(wing area), which is effectively (wing span)/(mean chord of wing).) The relationship between size and aspect ratio differs between passeriforms, on the one hand, and shore birds and ducks on the other. In the former, it appears that the wing becomes somewhat less dense as its area increases, i.e. there is no apparent aerodynamic requirement for greater structural strength as size increases. For shore birds and ducks, however, there is structural reinforcement as size increases, presumably because of the greater flying speed of the latter. If any of these relationships are optimal, their differences may reflect the different possible optimal types of flight, e.g. minimal energy cost per unit distance, minimal energy cost per unit time or maximal manoeuvrability (Pyke, 1981).

Using standard dynamic methods, Greenewalt concluded that the power required for flight is given by

$$P_r = \frac{\varrho}{2}(C_{D_p}S)\,V^3 + \frac{2}{\varrho}\,\frac{W^2}{\pi b^2 f}\,\frac{1}{V},$$

or if we wish to discuss "specific power", that is to say power per unit weight,

$$P_r/W = \frac{\varrho C_{D_p}}{2}\frac{S}{W}V^3 + \frac{2}{\varrho}\,\frac{W}{\pi b^2 f}\,\frac{1}{V}.$$

Here, W is the mass multiplied by $\mathbf{g}$, V is the velocity, ϱ is the density of air, S is the projected wing area, C_{D_p} is the parasitic drag coefficient for the entire organism, b is the wing span, f is an efficiency factor and P_r is the power.

From this analysis, we can conclude that a 1% increase in wing area requires a 1·3% increase in weight, which might not necessarily be advantageous. However, if we consider the power for flight per unit weight, we find that this is given approximately by

$$K_1\frac{1\cdot01}{1\cdot013} + K_2\frac{1\cdot013}{(1\cdot01)^2}$$

$$= 0\cdot997K_1 + 0\cdot993K_2.$$

In other words, a 1% increase in wing area gives approximately 0·5% less power for flight per unit weight. Thus, a small increment in wing area, other things being equal, may be expected to be advantageous.

This result is for an existing, highly evolved wing. If we consider a 1% increase in surface area for a falling body, i.e. the initial state of a leaping organism with little or no aerodynamic lift, then we find that a small increase in surface area allows a similar proportional distance per leap for the same expenditure of energy. The initial stages of evolution of gliding flight are thereby clearly explicable in terms of small advantage. This would not have surprised Darwin (1859), since he had made the same point without recourse to any aerodynamic argument whatsoever (cf. Fisher, 1954).

TOPICS FOR DISCUSSION

1. "Experimental tests of life history theory are not yet feasible." (Bell, 1980).
   Suggest minimal criteria for such tests to be possible.
2. "The possibility of life as we know it evolving in the Universe depends on the values of a few basic physical constants – and is in some respects remarkably sensitive to their numerical values." (Carr and Rees, 1979).
   This is what would be expected if life had evolved by natural selection, but is it also what would be expected if all evolution occurred by chance?
3. Thompson et al. (1980) "measured the cost of locomotion for three bipedal hopping eutherians (0·03–3·0 kg) and one bipedal hopping marsupial (1·1 kg). When properly trained for sustained running, none of the three species deviated

from the expected quadrupedal pattern relating cost of locomotion to running speed, even while running fully bipedally. Bipedality must therefore be associated with benefits other than those related to the energetic costs of locomotion.''

Suggest hypotheses which can be tested for these benefits in terms of traits which may be subject to natural selection.

4   early evolution

Darwin (1859) was one of the first to think of order arising in the primeval soup, and since then many have speculated on the constitution of that soup and how it might have changed, first simply by chemical means, and later by a series of self-reinforcing reactions which gradually changed into life. That is, at most stages, the presence or absence of life would have been a point debatable by disinterested observers. The nature of the initial freak of chance which brought about the possibility of life may never be properly identified, but on the basis of considerable experimental work, we can approach a discussion of very early evolution after that accident.

In the words of Eigen and Schuster (1979),

> The breakthrough in molecular evolution must have been brought about by an integration of several self-reproducing units to a co-operative system and . . . a mechanism capable of such an integration can be provided only by the class of hypercycles.

A hypercycle is a set of interlocking feedback loops and in terms of early evolution requires primitive forms of both copying and translation (cf. Eigen, 1971; Eigen and Schuster, 1979; Eigen *et al.*, 1980). These must allow highly specific coupling of protogenes and proto-enzymes, and White (1980) has suggested that early evolution is far more likely to have been very imprecise, and has suggested the following requirements for what he has called an autogen model, an autogen being two short oligonucleotide sequences coding for two simple catalytic peptides. Requirements for such a model are shown in Table 4.1.

This model has the implication that the products of the auto-catalytic process will be highly variable; errors will be extremely probable and, as White has noted, short regions could be propagated through longer chains, though varying as this happened, so that a large variety of related but not identical oligopeptides would be produced. Barbieri (1981) has taken this type of model a step further, proposing that the origin of life resided in the translation apparatus, so that RNA rather than DNA would be the primordial genetical material. This would ease constraint 4 in Table 4.1. Barbieri proposes that DNA subsequently became the primary information storage and transmission medium through its superior stability; the mech-

Table 4.1 Requirements and assumptions of the autogen model.

1. Steady or periodic availability of free energy and monomers (amino acids and nucleotides)
2. Low-yield synthesis of oligopeptides and oligonucleotides
3. Kinetic stability of oligomers
4. Localization or partial containment of oligomers. Products of replication and translation remain near the parent oligonucleotide much of the time
5. Crude oligonucleotide selectivity of amino acids during oligopeptide synthesis (primitive translation process of unspecified mechanism)
6. Crude oligonucleotide replication to produce complementary oligonucleotides
7. Two short peptides or families of peptide sequences which catalyse replication and translation without recognition or selectivity toward any particular oligonucleotide sequence

From White (1980).

anism is unspecified. The finding that transfer RNA has changed very rapidly over evolutionary time and the suggestion that some tRNAs may have independent origins in phylogeny (Cedergren *et al.*, 1980) might be regarded as evidence against Barbieri's hypothesis, in that it suggests that the translation apparatus was at least extensively modified much later than the time when DNA became the primary medium.

Many studies (see e.g. Lohrmann and Orgel, 1979, and Nelsestuen, 1980, for references) have confirmed the first three requirements in Table 4.1. The last three are as yet less certain, and the fourth, as noted, is problematic. However, it is of great interest that the possibility mentioned above, namely that particular sequences might be propagated through the primordial genome, appears to have occurred in bacterial evolution (Ornston and Yeh, 1979). It appears that at replication misalignment of DNA in one region may affect replication elsewhere, so that substitution of copies of DNA sequences into structural genes may be a mechanism for information-increase more often than tandem duplication. Furthermore, the same change can occur preferentially, allowing rapid responses to environmental change "by bacterial genomes that are remarkably resistant to genetic drift" (Ornston and Yeh, 1979). While this mechanism does not appear to be the norm in eukaryotes there is some evidence that it may have occurred in ribosomal gene sequences in higher eukaryotes such as primates (Arnheim *et al.*, 1980) and perhaps even in the much less highly repeated α-globin loci of man (Liebhaber *et al.*, 1981).

Thus, there is considerable indirect evidence for the concept that early evolution might have occurred by a self-accelerating process, traces of which may still be seen. Genome evolution by such mechanisms might well go on, even without phenotypic advantage to the organism, until some phenotypic disadvantage arose (cf. e.g. Southern, 1970; Orgel and Crick, 1980; Doolittle and Sapienza, 1980). According to Ohta and Kimura (1981), the variance in the amount of functionless DNA will depend on both the rates of duplication and deletion, and the rate of chromosomal mutation, even when there is no systematic change in the mean quantity of such DNA. If this DNA is available for subsequent change into information-carrying DNA, then chance will be important at every stage. At a higher level of organization, selection might be accelerated through the joint selection of the phenotypes resulting from highly organized sets of genetic material such as single chromosomes (Mather, 1949, Franklin and Lewontin, 1970; Slatkin, 1972). However, this latter topic is not relevant here (cf. Section 6.2.1).

As a final example of a still-existing process which might have been influential in early evolution, consider random-in-time mating. This was the process suggested by Visconti and Delbrück (1953) to explain the occurrence of triparental phage particles following infection of bacteria by three different phage types. Its properties as a regular process have not been much investigated (Bennett, 1954; Mayo, 1967), but under the conditions envisaged by Eigen and White, it may well have been the norm. That is, multiple, frequent recombination, both regular and irregular, may have yielded new combinations much more rapidly than in a regular, highly developed system of meiosis and zygosis. Possibly the genetics of subcellular organelles such as mitochondria is similar (Thrailkill *et al.*, 1980; Takahata and Maruyama, 1981).

TOPICS FOR DISCUSSION

1. What kinds of evidence will allow one to distinguish between the models of Eigen and White?
2. "It is believed that the Earth's magnetic field may have been one of the main factors the combined effects of which have led to the dissymmetry of living beings. A magnetic field has a plane of symmetry perpendicular to the magnetic lines of force. Therefore it exerts a dissymmetric force only in conjunction with some physical phenomenon the effect of which is non-symmetrical with respect to the plane of symmetry." (Gladyshev and Khasanov, 1981).

   Would this hypothetical effect of magnetism be expected to occur only with natural selection or in any mode of evolution?

5 adaptation

Adaptations have long been regarded essentially as "adjustments in the organism to its environment" (Eigenmann, 1909). The sense in which adaptation is regarded here was first described by Fisher (1930*b*).

> An organism is regarded as adapted to a particular situation, or to the totality of situations which constitute its environment, only in so far as we can imagine an assemblage of slightly different situations, or environments, to which the animal would be on the whole less well adapted; and equally only in so far as we can imagine an assemblage of slightly different organic forms, which would be less well adapted to that environment. This I take to be the meaning which the word is intended to convey, apart altogether from the question whether organisms really are adapted to their environments, or whether the structures and instincts to which the term has been applied are rightly so described.

Fisher was then able to show that:

> If therefore an organism be really in any high degree adapted to the place it fills in its environment, this adaptation will be constantly menaced by any un-directed agencies liable to cause changes to either party in the adaptation. The case of large mutations to the organism may first be considered, since their consequences in this connexion are of an extremely simple character. A considerable number of such mutations have now been observed, and these are, I believe, without exception, either definitely pathological (most often lethal) in their effects, or with high probability to be regarded as deleterious in the wild state. This is merely what would be expected on the view, which was regarded as obvious by the older naturalists, and I believe by all who have studied wild animals, that organisms in general are, in fact, marvellously and intricately adapted, both in their internal mechanisms, and in their relations to external nature.
>
> Such large mutations occurring in the natural state would be unfavourable to survival, and as soon as the numbers affected attain a certain small proportion in the whole population, an equilibrium must be established in which the rate of elimination is equal to the rate of mutation. To put the matter in another way we may say that each mutation of this kind is allowed to contribute exactly so much to the genetic variance of fitness in the species as will provide a rate of improvement equivalent to the rate of deterioration caused by the continual occurrence of the mutation.
>
> As to the physical environment, geological and climatological changes must always be slowly in progress, and these, though possibly beneficial to some few organisms, must as they continue become harmful to the greater number, for

the same reasons as mutations in the organism itself will generally be harmful. For the majority of organisms, therefore, the physical environment may be regarded as constantly deteriorating, whether the climate, for example, is becoming warmer or cooler, moister or drier, and this will tend, in the majority of species, constantly to lower the average value of m, the Malthusian parameter of the population increase. Probably more important than the changes in climate will be the evolutionary changes in progress in associated organisms. As each organism increases in fitness, so will its enemies and competitors increase in fitness; and this will have the same effect, perhaps in a much more important degree, in impairing the environment, from the point of view of each organism concerned. Against the action of Natural Selection in constantly increasing the fitness of every organism, at a rate equal to the genetic variance in fitness which that population maintains, is to be set off the very considerable item of the deterioration of its inorganic and organic environment. This at least is the conclusion which follows from the view that organisms are very highly adapted. Alternatively, we may infer that the organic world in general must tend to acquire just that level of adaptation at which the deterioration of the environment is in some species greater, though in some less, than the rate of improvement by Natural Selection, so as to maintain the general level of adaptation nearly constant.

An increase in numbers of any organism will impair its environment in a manner analogous to, and more surely than, an increase in the numbers or efficiency of its competitors.

The essential features of this argument are, first, that a high degree of adaptation is the rule in nature. This comes from observation. For example, as Barlow (1981) has pointed out for human eyesight,

The highest resolving power achieved by man (nearly 60 cycle/deg) is quite close to the limiting resolution (D/λ cycle/radian) set by the diameter of his pupil (D, about $2 \cdot 5$ mm in bright light) and the wavelength of light used ($\lambda = 560$ nm).

As Barlow added, an obvious question raised is why the pupil should not be larger, but here we merely note the very high precision of adaptation, and the fact that this could still be higher. (Adaptation is used here solely in its evolutionary sense, as defined above.) Secondly, Fisher argued that the environment is constantly deteriorating. Thus, a genetical change observed may simply be that required to prevent things from getting any worse. Van Valen (1974) has proposed that environmental deterioration at a relatively constant rate may be the reason for the relatively constant rate of amino acid substitution in the evolution of proteins. It is not clear, however, that this substitution rate is constant, at least within limits which require such an explanation (Section 6.3).

An implication of Fisher's argument is that adaptation will rarely be optimal; if the optimum is constantly altering, the genotype will always lag in its approach to that optimum. How does this apply to an investigation of Orians and Pearson (1978), considered further by Lewontin (1978a, b),

Maynard Smith (1978) and Brady (1979)? The example is as follows (Brady, 1979):

> A researcher studies the behaviour of foraging birds. Since these birds carry the food back to the nest once they pick it up, were they to take the first item that they came upon the piece of food might be too small to make up for the energy lost in the round trip back to the nest and out again. Thus the researcher proposes that the choice of food particle, with regard to size, will not be random. Instead, the choice will represent an adaptation, and it will optimize the net energy gain of feeding. By surveying the actual distribution of food particles in the foraging area the researcher calculates the size that represents the optimal solution. It is not, of course, restricted to the largest food items, for these are too poorly distributed to be worth the energy needed to search them out. All this done, the researcher compares his figures with the bird's behaviour, and it turns out that the birds are biased in the direction of large particles, but not to the optimal particle size . . . The miss can be explained if we assume that another parameter is at work. The behaviour exhibited by the birds represents a compromise between the demands of energy efficiency and those of predator pressure – the birds cannot stay away from the nest long enough to conduct a proper search for the optimum particle size, since while they are gone their young are exposed to predators.

The problem here, very simply, is that an additional *ad hoc* hypothesis has been invoked (though Maynard Smith (1978) has suggested how this might in turn be tested and Herbers (1981) has considered in the context of colonies of ants how inept individual foraging may be consonant with efficient colony performance). An unlimited number of additional hypotheses could be used to explain the discrepancy between the original hypothesis and the observations, rendering the overall set of hypotheses irrefutable. However, it is evident that it is the original hypothesis which is at fault. Adaptation is a process which produces more or less highly adapted individuals, but only in a constant environment may we expect that the optimum will have been achieved. Definition of constancy is evidently a problem, but not a new one, and not one which can be ignored. Consider a slightly less contentious example, the interaction between human birth-weight and survival.

Human birth-weight is a character of complex determination. It is strongly influenced by a maternal genotypic effect (Robson, 1955; Morton, 1955). In addition, there are marked effects of maternal age and parity (Millis and Seng, 1954). Overall, it has been established by Hosemann (1948) and Karn and Penrose (1951) that the optimal survival of newborns is attained at a birth-weight substantially higher than the mean. For example, Fraccaro (1956) presented the following data for an Italian population: mean live birth-weight = 3291 ± 10 g, mean of all birth-weights = 3239 ± 11 g; optimum birth-weight was calculated to be 3611 g. Then the intensity of natural selection may be measured by Haldane's (1954) relationship $I =$

$\ln s_o - \ln s$ where s_o is the optimal survival rate and S the population survival rate, giving $I = \ln 0 \cdot 969 - \ln 0 \cdot 933 = 0 \cdot 038$. Ulizzi *et al.* (1981) have shown that most of the directional selection occurs for premature babies, the optimum being much further above the mean than is the case for full-term babies, but both stabilizing and directional selection occur for all birth ages.

Here we know that the environment has changed rapidly; unusually, however, the change in the environment has been an improvement, leading, for example, to a marked reduction in the variance in birth-weight in Italy (Terrenato *et al.*, 1981). One does not need to speculate about the optimum birth-weight in a different environment to realize that one would expect it to be different, given the strong selection now seen for birth-weight in relatively favourable conditions. If, however, one had expected the adaptation to be complete, i.e. the trait to be at its optimum, then one would be surprised in the manner of Brady (cf. also Maynard Smith, 1978).

In the reed bunting, *Emberiza schoeniclus*, where, in contrast to man, a relatively constant environment appears to have persisted for some time in one population in Finland, the modal clutch size is also the most productive. in terms of survival to reproductive age (Haukioja, 1970). In the magpie, *Pica pica*, Högstedt (1980) has obtained evidence for adaptation at least as precise, since different distributions are found in different regions with the modal size being optimal in each case.

We see, therefore, that adaptation may be extremely precise, but that there is no logical necessity for it to be optimal in any given case. Two questions naturally arise: first, is there any reason to suppose that all directional genetical change is optimizing, and secondly, is the mechanism precise enough to achieve an optimum where departures from it are extremely small?

5.1 NON-ADAPTIVE DIRECTIONAL CHANGE

It was recognized quite soon in the development of evolutionary genetics that fitness will not necessarily always be maximized, even in a constant external environment. Slatkin (1978) has considered the alternative possibility, that fitnesses will be equalized approximately in a population which is polymorphic. Fisher (1941) was, however, the first to show clearly how the population mean fitness would not necessarily rise under natural selection, so it is worth presenting his argument in some detail.

> Early selectionists, following in this respect the language of earlier theological writers on organic adaptation, often speak of selection as directed "for the good of the species". In reality it is always directed to the good, as measured by descendants, of the individual. Unless individual advantage can be shown,

natural selection affords no explanation of structures or instincts which appear to be beneficial to the species. Yet in Wright's equation

$$\left[\Delta p = \frac{pq}{2W} \frac{d\overline{W}}{dp} \right]$$

the whole evolutionary sequence would appear to be governed by the principle of increasing the "general good". It may therefore be worthwhile to examine in detail a model involving powerful selection, in which the fitness of the species as a whole, judged by external criteria, is entirely inoperative.

The "good of the species" is still invoked as an evolutionary argument (cf. e.g. Baldwin and Krebs, 1981), but we are here concerned, in Fisher's model, with the dynamics of a gene which ensures self-fertilization without affecting male or female fertility, as might happen by cleistogamy. Fisher used the following model:

Genotype	Fertilization
gg	open pollinated
Gg	half open pollinated half selfed
GG	all selfed

It can be shown that the gene G always increases in frequency. As Fisher noted, this increase comes about regardless of the environment, and of the relevance of self-fertilization to the current state of the environment, apart from any selective pressures such as those considered below on problems associated with inbreeding depression.

A gene which increased rapidly in frequency to fixation would, on Fisher's view of the evolution of dominance, not be likely to be dominant or recessive, but one should also consider the case where it is either completely recessive or completely dominant. For example, if self-fertilization were completely recessive, we would have instead

$$R \quad gg \qquad \text{selfed}$$
$$2Q \quad Gg \; \Big\} \qquad \text{out-crossed}$$
$$P \quad GG \Big\}$$

$$p = P + Q$$

		GG	Gg	gg
$GG \times GG$	P^2	1		
$GG \times Gg$	$4PQ$	½	½	
$gg \times gg$	R^2			1
$Gg \times Gg$	$4Q^2$	¼	½	¼

$$p' = \frac{P^2 + 3PQ + 2Q^2}{P^2 + 4PQ + R + 4Q^2}$$

$$p' < p \text{ if } R^2 > 0.$$

Thus, once g has arisen by mutation, it always increases in frequency, and the same applies if self-fertilization is dominant. However, this assumes that all genotypes are otherwise selectively neutral. Thus, if we consider the case of cleistogamy mentioned by Fisher, we might expect that an out-pollinated organism would have undergone selection so that mutations affecting floral morphology were largely recessive, on account of their deleterious nature, so that when one ensuring complete self-fertilization arose, it might be expected to be recessive. Suppose that in such a case gg is at a selective disadvantage s through its inferior viability. Then g will increase in frequency only if

$$R^2 > s(Q^2 + R).$$

This means that when the mutant arises in a population its frequency will at first tend to decrease through selection, and will only increase in frequency if its frequency becomes elevated by random sampling so that the condition given holds.

5.2 PRECISION OF THE MECHANISM OF NATURAL SELECTION

As Fisher (1958) noted, '[i]t is inheritance that is discontinuous, not evolution'. This is the main limit to the precision of adaptation in one sense. But one may ask further whether there are thresholds, e.g. in perception, which may limit adaptation. Fisher (1930b) drew attention to the lack of evidence for such thresholds, and suggested that for very small differences in traits of adaptive significance selective factors associated with these differences would be approximately proportional to their magnitudes. Random processes would affect the perception of the differences, but the frequency of detection over a large number of occurrences nevertheless would be proportional to the magnitude of the differences.

These ideas have been tested by Duncan and Sheppard (1963, 1965).

Duncan and Sheppard (1963) confirmed Fisher's early suggestion that under conditions of random environmental variation very small changes in a stimulus are detectable on a statistical basis, and showed further that, as they had predicted theoretically, a threshold in perceptual ability could appear in an experiment as an artefact of the technique used. They concluded that there was no evidence for the existence of a threshold in discrimination.

Recent evidence on visual perception appears to support this view (Georgeson and Harris, 1981). In 1965, Duncan and Sheppard simulated Batesian mimicry in a laboratory experiment with domestic poultry as "predators" and green-coloured drinking solution as "prey". The darkest solution was the model, and when it was taken the bird was given an electric shock. No shock was given when a pale solution (the "mimic") was drunk. The results were in accord with the earlier work: the number of times a solution was drunk depended on the shock level received by the bird when it took the darkest solution, and was less at the higher shock level. At both shock levels the darker mimic solutions were taken less frequently, but the regressions of frequency of drinking on colour were different for the two levels of shock, showing that the darkest mimics were less readily distinguished with the higher shock. Their overall conclusions may be quoted:

> The relative selective advantage gained by a mimic is dependent on the penalty which accrues to the predator when it mistakenly takes the distasteful model which is being mimicked. When the consequences are severe, there is minimal selective advantage in improving mimicry beyond a certain point; however, if the consequences are mild, selective advantage continues to operate until a perfect resemblance is produced.

TOPICS FOR DISCUSSION

1. "Disruptive colour patterns, consisting of high contrast markings that serve to break up the outline of an organism, are among the classical types of protective coloration common in the animal kingdom . . . Few concepts in the theory of adaptive coloration are as well accepted, but as poorly documented, as that of disruptive coloration. No direct experimental tests demonstrating its efficacy have yet been performed." (Silberglied *et al.*, 1980).

   Devise a genetical model for the evolution of disruptive coloration by natural selection and use it to design an experiment to test the hypothesis that disruptive coloration cannot evolve in this way.

2. Green (1980) has developed a stochastic model for birds searching a patchy environment for prey, and has claimed that his "model shows that if patches vary in quality there are simple strategies of searching patches based on assessment of patch quality that enable a predator to do better than a predator that does not assess patch quality. The question remains of whether predators actually do use strategies that take advantage of variability in patch quality."

   Suggest how this question might be answered.

3. "That every theory of evolution must be consistent not merely with progressive development, but with indefinite persistence in the same condition and with retrogressive modification, is a point which I have insisted upon repeatedly from the year 1862 till now." (Huxley, 1895).

   What are the implications of this concept for our understanding of adaptation?

6 the rate of evolution

As Haldane (1954) wrote, "evolution is an almost unimaginably slow process", which is, "at any moment, almost in equilibrium." He listed ten factors responsible for this quasi-equilibrium: spatial clines, temporal clines, heterozygous advantage, sexuality and self-incompatibility factors, sex differences in genotypic fitness, mutation–selection balance, host–pathogen interactions, "non-Darwinian" selection, i.e. frequency-dependent selection, non-Mendelian inheritance and neutral genes. He might have said, with Milton (1671), "What boots it at one gate to make defence And at another to let in the foe?" and added Fisher's idea, discussed in the previous chapter, of the constantly deteriorating environment as an eleventh factor. Even without this, Haldane's formidable list means that several rather different topics are relevant to the question of evolutionary rates. Here, we shall first consider theoretical rates of change for well-defined model systems based on Mendelian inheritance. Such analyses may, though unfortunately they need not, indicate the limits to the rate of progress under natural selection.

Secondly, we shall consider actual rates of change, at the level of the gene, the gene product, the karotype and the trait. Rates of speciation and extinction will be considered later.

6.1 SINGLE GENE SELECTION

The biological fitness of an individual is the relative contribution of that individual to the ancestry of future generations, usually measured by the number of offspring surviving to reproductive age, relative to the population mean. It thus incorporates differences in both viability and fertility. Genotypic differences in fitness are a simple description of the action of natural selection at a particular time.

The following types of selection are possible for constant selective values:

Type	Dominance	Fitness of			$\Delta q (q = f(A_2))$
		A_1A_1	A_1A_2	A_2A_2	
(1) Against A_2	Absent	1	$1 - s_2/2$	$1 - s_2$	$-\tfrac{1}{2}s_2q(1-q)/(1-s_2q)$

Type	Dominance	Fitness of A_1A_1	A_1A_2	A_2A_2	$\Delta q\,(q = \mathrm{f}(A_2))$
(2) Against A_2	Incomplete	1	$1 - hs_2$	$1 - s_2$	$-\dfrac{s_2 q(1-q)(q + h(1-2q))}{1 - (2q(1-q)hs_2 + q^2 s_2)}$
(3) Against A_2 (recessive)	Present	1	1	$1 - s_2$	$-s_2 q^2 (1-q)/(1 - s_2 q^2)$
(4) Against A_1 (dominant)	Present	$1 - s_1$	$1 - s_1$	1	$s_1 q^2 (1-q)/(1 - s_1(1-q^2))$
(5) Against homozygotes	Absent	$1 - s_1$	1	$1 - s_2$	$\dfrac{q(1-q)(s_1(1-q) - s_2 q)}{1 - s_1(1-q)^2 - s_2 q^2}$

For various values of q, s_1 and s_2, we find that the rank order of the different rates of change is as shown in Table 6.1.

These results all relate to an infinitely large population. If the population size, N, is small, then the sampling variance of q, $q(1-q)/(2N)$, becomes an important influence on q, the process of gene frequency change being stochastic rather than deterministic. The relative influences of the different types of systematic change, however, do not change.

Thus, we see that almost any pattern of gene frequency change under selection is possible, from none to very rapid indeed. It is of interest to see what selective values have been estimated in natural populations. In some

Table 6.1   Rank order of absolute magnitude of selective changes in gene frequency.

s_1	s_2	h	q	Rank order				
0·01	0·01	0·01	0·05	1	5	2	4	3
			0·5	4	2	1	3	5
			0·95	4	3	2	5	1
		0·1	0·05	5	1	2	4	3
			0·5	4	2	1	3	5
			0·95	4	3	5	2	1
	0·1	0·01	0·05	1	2	3	5	4
			0·5	2	1	3	5	4
			0·95	2	3	5	1	4
		0·1	0·05	1	2	3	5	4
			0·5	2	1	3	5	4
			0·95	3	5	2	1	4
0·1	0·1	0·01	0·05	5	2	(1	4)*	3
			0·5	4	2	1	3	5
			0·95	4	2	3	5	1
		0·1	0·05	5	1	2	4	3
			0·5	4	2	1	3	5
			0·95	4	3	5	2	1

*Equal.

cases, these have been obtained from the viability and fertility of individuals of different genotypes, in others from change in gene frequency over time. Table 6.2 illustrates values obtained.

Evidently, very strong selection exists in natural populations. Even where the source of viability or fertility differences is unknown, such differences between genotypes may readily be detected (Haldane, 1962; Colgan, 1981). For the extremely widespread polymorphism of enzymes of intermediary metabolism, evidence of selection is as yet limited, other than the frequent disadvantage to homozygous null mutants, but some cases are beginning to be reported, e.g. an apparent advantage to a mixture of genotypes, as against a population of one genotype, for a malate dehydrogenase locus in *Drosophila pseudoobscura* (Tôsić and Ayala, 1980). Overall, however, such polymorphisms appear to approximate closely to the nearly neutral state (see Phillips and Mayo, 1981, for discussion).

Table 6.2 Selection coefficients in natural populations

Case	s_1	s_2	h	Trait	Organism	Source
(2)	0	0·33–0·5	0·5–0·6	Melanism*	*Biston betularia*	Haldane (1956)
(3)	0	0·33	0	Melanism*	*B. betularia*	Haldane (1924)
(4)	1	0	–	Epiloia	*Homo sapiens*	Crow (1961)
(4)	0·33	0	–	Myotonic dystrophy	*H. sapiens*	Crow (1961)
(5)	0·06	0·04	–	Colour pattern	*Paratettix texanus*	Fisher (1930*a*, 1939)
	0·2	0·75	–	Haemoglobin S ('sickle haemoglobin')	*H. sapiens*	Allison (1955)
(3)	0	0·84	–	Melanism†	*Selenia bilunaria*	Fisher (1933*b*)

* Melanism dominant or semi-dominant.
† Melanism recessive.

The evolution of melanism in environments affected industrially is a case of great interest. Brady (1979) has commented on these changes in cryptic colouration as follows: "[t]hat certain conditions can cause selective mortality means only that some alleles can be weeded out, not that this action can combine with variation in order to optimise adaptation." Now gene frequency change has been very rapid in the evolution of melanism (from $q \simeq 1-10^{-5}$ to $q \simeq 0$ in 100 generations; Haldane, 1956; Kettlewell, 1973). There has thus been very limited time for the new phenotype to be optimized. Despite this, dominance of the major gene has been modified, viability changes unrelated to visual predation have occurred, and pattern changes have also developed, all of which have tended to improve the new phenotype. Studies in progress (see e.g. Bishop, 1980) will elucidate the evolutionary significance of these changes and have already shown how unstable is the interaction between phenotype and environment. (And it is noteworthy that these changes would all be undetectable in the fossil record.)

6.2 RATE OF CHANGE IN FITNESS

Fitness, in the sense of ultimate reproductive success, is evidently only measurable after the fact, i.e. over particular known lineages for defined times. Indeed, it could be argued that it is only known precisely for extinct lineages or organisms which leave no descendants. None the less, as a relative concept it is simple to comprehend, though difficult to measure. For many purposes, its partition into viability and fertility is not even required, although the variances of viability and fertility in a population in a relatively constant environment constitute the boundaries within which natural selection can occur. Crow (1958) combined these into an index of opportunity for selection, the ratio of the variance in the number of progeny per parent to the square of the mean number of progeny per parent, i.e. essentially the departure of a population from the state where every individual reproduces itself precisely once. Since there are problems in interpreting the fertility component of the index (cf. e.g. Mayo *et al.*, 1978), it is not an absolute measure, and the results in Table 6.3 should be treated with caution.

Table 6.3 does, however, show how this index and its components may vary, even in the absence of a major change in the environment, such as colonization of a new habitat or the onset of an ice age. More precise comparisons between populations or species may usually be taken to be meaningless, since individual fitness is always relative to other members of a

Table 6.3 Index of opportunity for selection in different human populations.

	Population	Component		Total
		mortality	fertility	
Early	Xavante	0·49	0·41	0·90
	Yanomama	0·22	0·66	0·88
	Chile	1·38	0·17	1·78
Agricultural	Uganda	1·78	0·42	2·20
	Nyasaland	0·32	0·43	0·75
	Gold Coast	0·79	0·43	1·22
	Sweden (1820–90)	0·34	1·53	1·88
	Chile	0·33	0·22	0·62
Modern	USA	0·03	0·92	0·95
	Chile	0·15	0·45	0·67

Modified from Rasmuson (1980), Hed and Rasmuson (1981) and Crow (1966).

population in a certain environment. Changes in fitness over time within a population are of critical importance, for they display the course of adaptation of that population to its environment.

The Fundamental Theorem of Natural Selection (Fisher, 1930b), is still the best theoretical guide to the rate of change in fitness. We can assess it by considering a special case, Ewens's (1979b) single locus version for an infinite population:

$$
\begin{array}{cccc}
 & A_1A_1 & A_1A_2 & A_2A_2 \\
\text{frequency} & (1-q)^2 & 2(1-q)q & q^2 \\
\text{``fitness''} & \alpha & \beta & \gamma
\end{array}
$$
$$\text{mean fitness} = (1-q)^2\alpha + 2(1-q)q\beta + q^2\gamma$$

In the next generation this becomes

$$(1-q')^2\alpha + 2(1-q')q'\beta + q'^2\gamma$$

so that $\quad \Delta W = 2(1-q)q\,[\alpha(1-q) + \beta(1-2(1-q)) - \gamma q]^2.$

But this is the additive genetic variance in fitness, V_A, i.e. the genetic variance V_G less the components due to dominance, V_D,

$$V_A = V_G - V_D$$
$$= (1-q)^2\alpha^2 + 2(1-q)q\beta^2 + q\gamma^2 - ((1-q)^2\alpha + 2(1-q)q\beta + q^2\gamma)$$
$$- (2(1-q)q(\beta - (\alpha+\gamma)/2))^2.$$

V_A, V_G and V_D were defined by Fisher (1918) in his pioneering attempt to partition observed variability into components attributable to well-established causes. To make the problem soluble, he assumed that the genes affecting a trait acted additively, that they were all diallelic, that they were very numerous, and that they acted independently. Then the average effect of substituting any gene for its allele could be calculated, and that part of the genotype useful for predictive purposes between generations, the additive part, or the "essential genotypic" part, as Fisher called it, could be distinguished from that due to dominance. Their variances are V_A and V_D respectively. (Other complications to this very simple fundamental model need not be considered here; see Mayo et al. (1983) for discussion.)

Ewens's simple argument shows that for a single locus, the rate of change in fitness is given by the additive genetic variance in fitness.

Ewens (1979b) has extended this argument to show that the Fundamental Theorem of Natural Selection holds if and only if there is panmixia in the absence of migration, or fitnesses are solely additive (cf. also Samuelson 1978), i.e. during major gene frequency change, and there is approximate linkage equilibrium (small selective differences with loose linkage). (Selection of additive variation does not, therefore, depend on the build-up or break-down of linkage combinations, unless linkage is very close indeed.)

Now we want to know how the rate of change *varies* over time. Assume constant selective coefficients. Then if $V(p)$ is the variance in p over time,

$$V(\Delta W) \simeq \left(\frac{d(\Delta W)}{dq} \right)^2 V(q),$$

where

$$\frac{d(\Delta W)}{dq} = 2\alpha^2(3q^2 - 4q^3) + 2\beta^2(1 - 10q + 24q^2 - 16q^3)$$
$$+ 2\gamma^2(1 - q)^2(1 - 4q) + 4\alpha\beta q(2 - 9q + 8q^2)$$
$$- 8\alpha\gamma q(1 - q)(1 - 2q) - 4\beta\gamma(1 - q)(8q^2 - 7q + 1).$$

Comparing ΔW and $V(\Delta W)$, we obtain $\Delta W > > V(\Delta W)$ for all gene frequencies for frequently considered cases such as $N = 10, 100, 1000$ with the following viabilities:

	α	β	γ
deleterious recessive	1	1	0·9
"dominant" lethal	1	0·5	0
balanced polymorphism	0·5	1	0·5

As a corollary to the Fundamental Theorem of Natural Selection, Waddington and Lewontin (1967) have stated that

> any tendency to increase the quantity of information in the genome during evolution will be held in check because the rate of [evolutionary] advance will be inversely proportional to the number of information units,

because for a given range of variability of a metric trait the additive variance in a diploid is inversely proportional to the number of loci determining the character. Mayo (1971) showed that autopolyploidy might provide one way of escape from this tendency to stasis.

Other major modifications to the genome may have similar effects.

6.2.1 correlated change in a quantitative trait

The models considered above relate to change in fitness. Falconer (1966) and others (cf. e.g. Kempthorne, 1957; Demetrius, 1974; Crow and Nagylaki, 1976) have considered what is, in effect, the correlated response in a metric trait and this is examined further here. Actual rates of change for quantitative traits will be considered in Section 6.4. For a population with discrete generations, Lande (1976) obtained the following expression for the change in a phenotypic trait as a result of a change in fitness:

$$\Delta P(t) = h^2 V_P \frac{\delta \ln \overline{W}}{\delta P(t)},$$

where $\overline{W}$ is population mean fitness, $P(t)$ phenotypic value in generation t, h^2 heritability of the trait and V_p phenotypic variance. (Similar results have been obtained by Layzer (1978).) As Lande noted,

> this formula shows that with constant phenotypic fitnesses and infinite population size, the evolution of the average phenotype in response to selection is always in the direction which increases the mean fitness in the population.

In these terms, it is a corollary of the fundamental theorem of natural selection (cf. Robertson, 1968).

It is also important to note that selection of additive contributions to a trait may serve to increase complexity, and also to accelerate the rate of change. In attempting to quantify the arguments of Sewertzoff (1931) and Schmalhausen (1949), Layzer (1980) has argued that genes are of two kinds, those expressed in development (structural and regulatory genes) and those which regulate processes concerned with gametogenesis, especially recombination and mutation. If genes concerned solely with these latter processes exist, then, Layzer has suggested, they can act to increase variance in fitness during the early stages of a major phenotypic change and decrease it in the later stages, because such changes in variance are advantageous. This will come about by selection favouring increased mutation and recombination initially, decreased mutation and recombination later. This has also been suggested by Lande (1980a). Unfortunately, as pointed out by Templeton (1981), Layzer assumed an unrealistic symmetrical fitness function, so that these conclusions lack generality (cf. Gillespie, 1978). The assumption that the genes affect only mutation or recombination is also unrealistic, but even if the genes have other effects, the marginal fitness argument of Ewens and Thomson (1977) may mean that the selection process can still augment or diminish fitness variance to some extent, if the marginal effects of the genes in question are indeed those relating only to mutation or recombination.

A further implication of this type of argument is that functionally related developments may become increasingly related in ontogeny (as predicted by Sewertzoff (1931) and Schmalhausen (1949)). This will occur because mean fitness is proportional to the correlation coefficient for fitness determined by two characters having a continuous joint distribution. Hence, whether correlation is positive or negative, its strength will increase under selection. Whether the argument holds for highly skewed fitness distributions is very dubious (Templeton, 1981), but it suggests an approach needing further development.

6.2.2 substitutional load

In 1957, Haldane suggested that a major constraint on the rate of change under natural selection must be the cost, in terms of genetical deaths, of replacing individuals bearing the genes which were in the course of replacement. The argument can be seen most simply for a population segregating at one diallelic locus:

genotype	A_1A_1	A_1A_2	A_2A_2
fitness	$1 + s$	$1 + hs$	1
panmictic frequency	$(1 - q)^2$	$2(1 - q)q$	q^2

Set $h = 0 \cdot 5$. Then the mean fitness of the population is $1 + s(1 - q)$ and the substitutional load is given by the departure of the mean from the maximum possible fitness, i.e. about sq. Genetical deaths over a certain time (from t_1 to t_2) as a result of gene substitution are given by

$$= \int_{t_1}^{t_2} sq \mathrm{d}t$$

$$= 2 \log_e \left(\frac{1 - q_2}{1 - q_1} \right)$$

$$= -2 \log_e (1 - q_1) \text{ for fixation of } A_2.$$

For population size N, NL individuals must die for A_2 to be fixed, i.e. NL/T per generation.

If substitutions start n generations apart, T/n substitutions are occurring at any time and there are $(NL/T)/(T/n) = NL/n$ selective deaths per generation. Then, $L = 30$ implies that $n \simeq 300$ for $NL/n = 0 \cdot 1N$.

Accordingly, $(1 + (L/T))^{(T/n)}$ is the number of offspring the fittest individual must produce and this is huge for small values of n, such as are implied in Table 6.4, the true values being quite unknown. But the fittest individuals do not occur: Ewens (1979) pointed out that in a population of 10^5 the most fit individual likely to be met is twice as fit as the mean (i.e. must produce about two offspring compared with the mean of one necessary to maintain constant population size).

Thus the cost of natural selection is not a constraint on the rate at which it can modify an organism.

This is a negative argument: consider a constructive alternative. Sved (1968) attempted to calculate the maximum possible rate of gene substitution in evolution using a probit model. That is, gene substitutions influence an individual's score on some hypothetical metric scale x with mean μ and variance σ^2 and an approximately Gaussian distribution, thus:

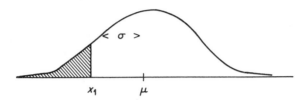

Then the selective loss of individuals having metric value x_1 is simply the hatched area shown, i.e.

$$\int_{-\infty}^{x_1} \frac{1}{\sigma\sqrt{(2\pi)}} \exp \frac{-(x-\mu)^2}{\sigma^2} \, dx.$$

Sved has shown (cf. the Fundamental Theorem of Natural Selection) that σ^2 and that proportion of σ^2 due to gene substitutions are the critical variables, and that quite large selective differences between genotypes for a given locus are possible without imposing any load on the population, providing the fundamental assumption is made that "the mean fitness of the population is not affected by its genotype". There is then no upper limit to the rate of gene substitution. One important corollary is that "if evolution occurs . . . as a multistage rather than a continuous process, it [may] no longer be true that the rate of change of gene frequency at individual loci is not important". The model thus accommodates both classical gene substitution under strong selection and phenotype modification through such processes as regulatory changes and genome reorganization.

The process considered in this chapter is in marked contrast to selection in the very early evolution of life (Chapter 4). Under the circumstances then hypothesized, with no dosage compensation (i.e. no dominance, no gene inactivation, etc.), selection pressures may be huge, and as few individuals may be expected to be well adapted, progress may be expected to be disorderly (non-Haldanian) and potentially very rapid.

6.3 RATE OF AMINO ACID SUBSTITUTION

The approach of Kimura (1968; but as set out in Kimura and Ohta, 1971) is the simplest and clearest. He suggested that this rate should be measured through comparison of the amino acid sequences in homologous proteins of related organisms, obtaining time since divergence from palaeontology. If n_{aa} is the total length of the amino acid chain, and if the chain differs between groups at d_{aa} sites, then the mean number of substitutions per site over the evolutionary period in question is

$$K_{aa} = -\log_e (1 - f_d),$$

where $f_d = (d_{aa})/(n_{aa})$ (Zuckerkandl and Pauling, 1965). K_{aa} has variance

$$\frac{f_d}{(1 - f_d)\, n_{aa}}.$$

Then the rate of substitution per amino acid site per year can be obtained as

$$k_{aa} = K_{aa}/(2T),$$

where T is the number of years since the evolutionary divergence from the common ancestor.

It can be shown that if mutant substitution is largely the result of natural selection, then the rate of substitution is approximately $4N_e s\mu'$; N_e is the effective population number, s is the selective advantage of the mutants and μ' is the mutation rate per gamete for such mutants. This will be the case if $4N_e s \gg 1$. If these substitutions are largely the result of random fixation of selectively neutral mutants, the rate of substitution is simply μ', the mutation rate per gamete for such mutants, probably $< 10^{-5}x$, where x is the number of structural genes per gamete (Neel and Rothman, 1981; Chakraborty, 1981).

In 1968, Kimura suggested that the evolutionary rate of change in haemoglobin chains was approximately $1 \cdot 0$ changes per 10^9 years. In 1971, Kimura and Ohta suggested approximately $0 \cdot 8$. Table 6.4 shows Kimura's estimates in 1977.

The mean rate in his earlier paper (Kimura, 1968) was approximately $\bar{k}_{aa} = 0 \cdot 5 \pm 0 \cdot 4$, while for the data in Table 6.4 it is approximately $\bar{k}_{aa} = 2 \cdot 1 \pm 3 \cdot 0$. It is of some interest that the rates, both for individual proteins and overall, appear to change with the increasing information available over time. Although the estimates are very imprecise, as indicated by the large standard errors, one is tempted to compare the changes with those reported by Youden (1972) for fifteen successive values of the astronomical unit, i.e. the average distance to the sun, obtained between 1895 and 1961. Each estimate lay outside the confidence interval of the one immediately preceding it and in twelve cases successive intervals did not overlap. As Williams (1978) noted in commenting on this, "The conclusion of systematic bias seems irresistible", the bias of course being in the procedures used, and Goodman (1981) has indeed claimed that the k_{aa} measures, like related

Table 6.4 Evolutionary rate in terms of amino
acid substition (per 10^9 years).

Proteins	k_{aa}	k_{nuc}
Fibrinopeptides	4–9·0	1·8–4
Pancreatic ribonuclease	3·3	
Haemoglobin chains	1·4	2·3 ± 1·1
Myoglobin	1·3	0·2–1·1*
Animal lysozyme	1·0	
Insulin	0·4	
Cytochrome c	0·3	
α-crystallin A chain	0·3†	
Histone IV	0·006	3·7 ± 1·4

Modified from Kimura (1977).
* From Goodman *et al.* (1975).
† From de Jong *et al.* (1977).

measures of his own (cf. also Baba *et al.*, 1981), are consistent under-
estimates of the actual unknown rates (but see Kimura (1982*a, b*) for
criticism of this argument).

Table 6.4 also shows some estimates of the rate of nucleotide change (cf.
also Kimura 1980*b*). These may be expected to be higher, because of the
degeneracy of the genetic code (Jukes, 1980), which is great enough to allow
even doubly degenerate codons in overlapping genes (Siegel and Fitch,
1980). That is, there are several cases where in overlapping genes codons are
degenerate (can change without altering the amino acid coded for) in both
genes. Overlapping genes would be expected to impose constraints on the
form and direction of evolution; degeneracy may be one partial method of
avoiding major constraints. (Degeneracy also has the implication that the
information content of DNA or RNA is significantly higher than that of
protein, so that information flow from protein to nucleic acid should
not occur – the "central dogma" – cf. Yockey, 1974, 1978, 1979). On the
other hand, it has been shown that the possible changes within a gene have,
over the substantial number of cases examined, a distribution very far from
what would have been expected if chance had been the sole determinant
(Holmquist and Pearl, 1980). This is because as successive mutations occur,
they alter what subsequent mutations may change either advantageously or
without effect.

Before considering the role of natural selection in these changes, we
should consider one other relevant point, namely the rate of change for any
particular protein. The implication of the work of Kimura (1968, 1977) and
many others (e.g. Wilson *et al.*, 1977), illustrated in Table 6.4, is that each
polypeptide chain has its own particular rate of change over time, with
stochastic fluctuations about this rate. Despite these fluctuations, it is sug-

gested that the rate is a useful measure of evolutionary change, especially if most of the divergence in the protein has been the result of fixation of neutral mutants. As Carlson *et al.* put it: "Protein clocks appear to have the potential for giving us a temporal view of evolutionary relations among all living species. By contrast, the fossil record is too poor to give us such a complete view." Nucleic acid clocks may give different results again (Ferris *et al.*, 1981).

If the method is to be widely used, one must therefore assess its variability. It is regarded by proponents of the clock method as reasonable that the clock time should have variability about two to four times that expected for the Poisson distribution (Nei, 1977; Fitch, 1976). Korey (1981) has shown theoretically that the transient distribution of amino acid variants, which is what is used to compare taxa, depends substantially on both population size and generation length, so that arguments based implicitly on the constancy of both will be subject to large, but unknown bias. Furthermore, some observations suggest that variability may be much greater than this while close examination of Nei's (1977) analysis suggests that he may have underestimated the variability substantially. For example, Lessios (1979), examining sea-urchins taken from both sides of the isthmus of Panama, has suggested that 20-fold differences in rates are possible in this reasonably closely related group, a divergence far greater than is allowable on the hypothesis of neutrality or near-neutrality. Again, van den Berg and Beintema (1975) have suggested, from an examination of pancreatic ribonucleases of hystricomorph rodents, that while the rate for this group of proteins shown in Table 6.4 is appropriate over a range of other organisms, from rat to giraffe, this is not so in these hystricomorphs. Divergence in the guinea-pig (*Cavia porcellus*), for example, has been at least 10-fold higher than the mean rate for the group, which is in turn substantially higher than the rate shown in Table 6.4. Wriston (1981) has suggested that since some, though not all, of the changes seen in the hystricomorphs are also seen in some New World monkeys they may represent convergent evolution. (Convergent evolution is most unlikely to occur by chance, but no selective explanation has yet been advanced.) Finally, Avise *et al.* (1980a, b) have produced evidence to suggest that New World wood warblers (*Paralidae*) have a "clock" which runs much slower than that in New World rodents (*Cricetinae*) and that North American sparrows (*Emberizidae*) have similarly changed very slowly.

We now return to the question of natural selection. As I showed earlier (Mayo, 1970), on Kimura's (1968) argument, large numbers of slightly deleterious mutants might be expected to be fixed by chance, but there is little evidence that this has been the case (though see Ohta, 1973, 1974, for an elaboration of this argument). Thus, the problem is essentially that of determining the relative proportions of actual changes which have occurred as the result of random fixation of neutral (or perhaps very slightly deleterious) mutants or as the result of natural selection of advantageous mutants. Ewens (1979a) has taken the view that null hypothesis of "strict neutrality", i.e. all new mutants are selectively identical, is of little interest and is no longer tenable. Not all classes of mutants occur equally frequently either; e.g. deletions appear to be more frequent than insertions, probably because of the way the double helix is repaired (de Jong and Rydén, 1981). However, use of this hypothesis did allow Corbin and Uzzell (1970) to estimate the proportion of all mutations which are deleterious. They began by considering the molecule or set of molecules which appeared to have evolved most rapidly, i.e. the fibrinopeptides, which are cleaved from fibrinogen in the process of blood clotting, and thereafter have no known function. They assumed that all possible mutations in fibrinopeptides are neutral, while all evolutionary substitutions in other proteins are neutral. They then found that the relative time for a substitution in fibrinopeptide chains as against others was 111 to 540, and therefore concluded that the proportion of deleterious mutations in non-fibrinopeptides would be at least $1 - (111/540)$, i.e. 80% or more.

Boyer *et al.* (1978) criticized Corbin and Uzzell's method, because two separate hypotheses were needed to produce the estimates, and carried out a very thorough investigation of the proportion of all point mutations in haemoglobin sequences which are deleterious. They considered 156 known mutations in the human β-haemoglobin chain, and classified them according to whether they were harmful or relatively benign. Six were unclassifiable, and the properties of the remaining 150 are shown in Table 6.5.

It should be noted that most of the mutants classified as benign have not been seen in the homozygote, where they might well be deleterious. Thus, the classification is conservative in terms of the proportion of mutants which are deleterious. The one definitely harmful mutant known to have an equivalent substitution in mammalian evolution, Haemoglobin San Diego,

Table 6.5 General properties of 150 beta haemoglobin variants.

	Harmful	"Benign"
Total number	86	64
Occurs at same position as a change seen during mammalian evolution	15	54
Represents same change as seen in mammalian evolution	1	14

From Boyer *et al.* (1978).

where the change is $\beta 109$ (Val → Met), is one of the haemoglobins with increased oxygen affinity (Bellingham, 1976), so that an evolutionary advantage or disadvantage might well be possible. Overall, the results in Table 6.5 could be interpreted as lending support to the neutral hypothesis in a less than strict form, but the further analyses of Boyer *et al.* (1978) suggest that at least 95% of all amino acid substitutions, i.e. non-synonymous mutations, are functionally unacceptable as homozygotes.

The results of investigations of a number of proteins tend to suggest that the three-dimensional structure of a protein is much more closely constrained than the sequence (Blundell *et al.*, 1981), even for taxa as widely separated as *Gallus domesticus* and bacteriophage T4 (Matthews *et al.*, 1981). For example, it might be that the changes which in human haemoglobin are "benign" in the sense of Boyer *et al.* (1978) are benign because they do not alter the three-dimensional conformation of the protein. However, adjacent or even more distinct changes, individually benign, may jointly be deleterious, or changes individually deleterious may be jointly less so (cf. e.g. haemoglobin C Harlem, which is "sickle haemoglobin" plus a second mutation; Bookchin *et al.*, 1970). It is of some note that the coding sequences separated by introns in the haemoglobin genes do not correspond to the structurally critical domains of these proteins (Rashin, 1981). However, it is not clear to what extent, if any, this is related to temporal change in the globin genes.

Peetz (1979) examined five of the classes of protein in Table 6.4 to ascertain whether charge changes occurring in evolutionary divergence have been random (and therefore possibly neutral) or not. He found that myoglobin has accumulated charge changes at about half the rate predicted by a model of random accumulation, while one of the fibrinopeptides has changed at about twice the rate expected in the random model. The other three proteins (cytochrome c, haemoglobin and insulin) show changes compatible with random accumulation, apart from the α haemoglobin of carp, which is anomalous (the question of polyploidy in the carp lineage may be related). Taking these results with those of Boyer *et al.* (1978), the range of possible neutral variation, though still substantial, is narrowed further by the clear suggestion of non-deleterious, non-random change.

Demonstration of actual evolutionary advantage would seem to require the occurrence in some experimental organism of a "back" mutation of one of the fifty-four benign evolutionary changes mentioned in Table 6.5. In such a case, careful investigation might reveal selectively relevant differences in properties between the two haemoglobin β chains. Such direct evidence will be very difficult to obtain.

Some indirect evidence comes from globin "pseudogenes", i.e. nucleic acid sequences closely homologous to the globin structural genes, but not expressed, presumably because of particular mutations. These sequences have evolved more rapidly, assuming them to have originated with the duplication of functional globin genes, than those functional genes (Kimura, 1980b; Li et al., 1981). The simplest explanation for this is that the additional changes represent substitutions which have not been selected against because the pseudogene is functionless. Thus, this may be evidence for neutral change in the pseudogene and stabilizing selection on the functional gene.

6.4 RATE OF CHANGE IN A QUANTITATIVE TRAIT

To assess changes over evolutionary time one depends on the fossil record, but it is far from clear how perfect this is. For example, pterosaurs, having delicate skeletons relative to more robust animals are unlikely to have been preserved as fossils, yet they have been discovered in every continent (Bramwell and Whitfield, 1974). Furthermore, although the first known discovery of a pterosaur fossil was in 1784 (Seeley, 1870; Watson, 1974), it was only in 1980 that one was found in Australia (Molnar and Thulborn, 1980). To take another example, an entirely new subclass of extinct birds, distinct from existing birds and the two known extinct subclasses, has been described recently by Walker (1981) on the basis of Cretaceous fossils from Argentina. Finally, it is well known how fossils of the coelocanth were found early last century in strata dating from the Devonian to the Cretaceous, yet this century living examples were found (see Locket, 1980, for references). Thus, the patchiness or otherwise of the fossil record will reflect both the biases due to differences between animal or plant types and the intensity and extent of investigation.

Nevertheless, substantial data exist, and Haldane (1949) first discussed how one might quantitatively measure the rate of change of metrical characters. He gave two possible measures, the proportional rate of change in the average phenotype,

$$\frac{\log_e \bar{x}_2 - \log_e \bar{x}_1}{t},$$

and the rate of change in terms of the phenotypic standard deviation $\sqrt{(V_P)}$ as a unit, $\bar{x}_1$ and $\bar{x}_2$ being the mean values at the beginning and end of a time period of length t. He presented the results for the increases in length of six suborders of Dinosaur during the Mesozoic era (225–70 million years BP) which are shown in Table 6.6.

These annual changes are very small, as they are in the evolution of horses, which Haldane also considered. In general, Haldane concluded that the slower rates could have been the result of mutation pressure and that random effects might also have been important.

Table 6.6 The increase in length of six suborders of dinosaurs during the Mesozoic era.

Suborder	Annual rate	Time (years) ($\times 10^6$)
Sauropoda	$3\cdot0 \times 10^{-8}$	35
Stegosauria	$2\cdot1 \times 10^{-8}$	30
Theropoda	$2\cdot0 \times 10^{-8}$	97
Ornithopoda	$2\cdot2 \times 10^{-8}$	60
Ankylosauria	$0\cdot26 \times 10^{-8}$	40
Ceratopsia	$6\cdot1 \times 10^{-8}$	22

From Haldane (1949).

Mutation pressure is a plausible explanation for the degeneration over time of organs which are not in use. This is not to say that natural selection would be irrelevant; e.g. in the evolution of burrowing animals such as the mole, *Talpa europea*, or the marsupial equivalent *Notoryctes typhlops*, diminished or totally covered eyes might have been advantageous, being less readily injured or infected. Overall, however, mutation pressure could always be at work.

The question of mutation pressure has been considered most extensively in relation to minor human defects such as colour-blindness, a set of Mendelian, mainly X-linked traits (Kalmus, 1965), and myopia, a classical polygenic trait (Sorsby and Fraser, 1964; Post, 1971). If one compares the frequencies of colour-blindness genes in "civilized" and "primitive" communities, the inferred mutation rates are an order of magnitude higher than those usually estimated. Overall, these results are in conflict with the conclusion of Kimura (1980a) that for an isolated small population, the time to fixation for a given mutant is, roughly, the reciprocal of the mutation rate (in generations). Thus, mutation pressure is a very slow process. Furthermore, while one sees how diminution of an organ or function could be advantageous in the sense that it reduced the developmental and metabolic resources devoted to ontogeny and maintenance of the organ or function, the same plausibility does not apply to an increase in size, whether of an organ or an organism. Other explanations need to be sought for the widespread agreement with Cope's Rule.

Lande (1976) extended Haldane's analyses considerably, using the corollary of the fundamental theorem of natural selection mentioned in Section 6.2.1 to assess how much selection would be necessary to bring about a given amount of change. Analysing the same data on horses as Haldane, Lande found a maximal minimum mortality rate to achieve the observed change under selection of two selective deaths per million individuals per generation, selection so weak as to be likely to be swamped by random effects.

Lande was led to the conclusion that random divergence between different adaptive zones might be very important in small populations ($N_e < 1000$), though later (Lande, 1980b) he implied that the magnitude of any such change would be modest. In the earlier study, he assumed V_P and h^2 constant in order to test the null hypothesis of selective neutrality and estimate its intensity if present. However, V_P is not constant over geological time and may well vary cyclically about a trend for quite long periods. Analyses of data on *Pelycodus* supplied by Gingerich (personal communication, 1981: cf. Gingerich and Simons, 1977; Gingerich, 1980) appear to confirm this; variability about the temporal change in a measure of tooth size showed a curvilinear pattern of change. Calculations in terms of a constant V_P will probably underestimate the influence of selection in direction change. (The problem of variable V_P will not be relevant for equivalent studies of short-term selection.)

Assuming that cyclical effects are unimportant, and, further, that the data available are a representative random sample (which will frequently be disputed; cf. Rosenzweig, 1977; Schankler, 1981), the question then arises how one might detect selection as against drift, mutation pressure and other non-selective phenomena. One set of tests, advocated by Lande (1976, 1977a, b, 1979), depends on the following argument.

With heritable variation of h^2V_P in a population of size N_e, a trait is assumed to be distributed normally with mean $\bar{z}$ and variance h^2V_P/N_e. Then after t generations it can be shown that the variance is expected to be h^2V_Pt/N_e. If n populations are available which have diverged from a common origin, then the variance between populations $s_{\bar{z}}^2$ may be compared with h^2V_Pt/N_e and on the hypothesis of no selection, $s_{\bar{z}}^2 N_e/(h^2V_Pt)$ will be distributed as $F(n-1, \infty)$. Lande (1977a) tested several sets of experimental data by this method and found evidence of selection on traits such as wing length in *Drosophila pseudoobscura* as a result of temperature differences.

For evolutionary divergence as seen in the fossil record, N_e and h^2 will be unavailable. Lande (1979) therefore considered the evolution of correlated traits, so as to combine experimental and historical data. He examined the allometric relationship between mammalian brain and body masses:

$$\text{mass}_{\text{brain}} = k \, \text{mass}_{\text{body}}^{\alpha},$$

where k and α are constant. Then selection on brain or body size will produce a correlated response in the other trait, and selection on both will be considered by ontogenetic and other factors. These factors are summarized in the phenotypic and genetical covariance structures of the population.

Investigation of these structures, Lande showed, may reveal the action of selection on quantitative traits. For example, principal component analysis might reveal sets of traits independently responsive to selection.

In the case of brain–body allometry, Lande showed that the values of α observed were in some cases very substantially greater than those expected on the hypothesis of random change. He therefore calculated that directional selection had occurred more for brain size than body size in most mammalian orders.

6.4.1 genetical variability and rate of change in a quantitative trait

From the fundamental theorem of natural selection and its corollary mentioned in Section 6.2.1, it is possible that a relationship exists between genetical variability for a trait and the rate of its change under natural selection. There are, however, at least two problems in evaluating this expectation. First, there may be genetical variability in a trait which is masked either by the form of the trait or by the method of ascertaining genetic variability (Mayo *et al.*, 1977, 1978). Secondly, there is some evidence that high levels of heterozygosity may be associated with lowered variability for metric traits (see Mayo *et al.*, 1980 for discussion). Accordingly, there is considerable interest in attempting to assess whether objectively determined variability, e.g. the level of heterozygosity in a population, is associated with the rate of change of a metric trait. Soulé (1976) has suggested that only populations of very large size may be expected to have very high heterozygosity, but for the moment we shall ignore this complicating factor.

A number of studies of quantitative traits had in fact shown a positive association between variation and rate of change (e.g. Farris, 1966, 1970; Kluge and Kerfoot, 1973), while contradictory views have been expressed by others (e.g. Bader, 1955; Long, 1969). Johnson and Mickevich (1977) therefore examined genetic variability by considering the information statistic

$$H = -\sum_{i=1}^{n} p_i \log p_i,$$

rather than heterozygosity. They obtained this for a large number of sets of isozymes in *Menidia*. They then obtained a significant, positive correlation between variability and evolutionary rate for quantitative traits and for allozyme traits. These results have been challenged by Riska (1979), on both methodological and genetical grounds, namely the sampling properties of H and also bias, through the inability to consider invariant traits.

6.5 RATE OF KAROTYPE EVOLUTION

Karotype evolution occurs when chromosomal mutations arise and reach high frequencies or become fixed in a population. (I consider polyploidy in Chapters 7 and 10.) The most important types of chromosomal mutation, in terms of karyotypic change, are inversions and translocations, including centric fission and fusion, though duplication may be more important as an initiator of evolutionary change.

When a chromosomal mutation occurs it must virtually always arise in heterozygous form. Meiosis will then yield unbalanced gametes which will cause zygotic wastage. Bengtsson and Bodmer (1976*b*) have shown that, in man, carriers of balanced reciprocal translocations have a fitness decrease of approximately 30%, while carriers of centric fusion mutations have fitness reduced by only about one-third as much. Minor chromosomal mutations, less easily detectable by existing methods, may have much smaller effects, but they will still be significant.

It might be expected that virtually all chromosomal mutations would therefore be lost from a population unless they possessed a very marked selective advantage to compensate for the effect on fertility. However, this is not necessarily the case. Bengtsson (1980*a*) has listed five major mechanisms for an increase in frequency of a chromosomal mutation.

(1) Segregation distortion: if, for some reason, the chromosomal heterozygotes produce more gametes bearing the chromosomal mutation than bearing the standard chromosomal complement, then the mutation may spread.
(2) Selective advantage: there may be a direct, selective advantage to the heterozygotes which compensates for the negative effect on fertility.
(3) Recombination modification: if, for example, an inversion effectively fixes a balanced, doubly heterozygous combination this may be a strong enough advantage to ensure the spread of the inversion.
(4) Inbreeding and homozygote advantage: under strong inbreeding, if the homozygotes for the chromosomal mutation are at an advantage to the normal homozygotes, then the mutation may spread regardless of the fitness of the heterozygotes.
(5) Random genetic drift: if population size is small enough, a chromosomal mutation may become fixed by chance, even if it is associated with a significant disadvantage.

The relative importance of these different mechanisms is as yet unknown. Strong selection is probably present in many cases of polymorphic chromosomal mutation (cf. Hartl *et al.*, 1980; Patton *et al.*, 1980). However, given that chromosomal mutations are very frequent (perhaps 2% of all live births in man, for example), the opportunity for chance effects to be very important is clear. Analysis of complex sex chromosome systems in metatherian systems which have probably arisen through chromosomal mutations (Martin and Hayman, 1966; Murtagh, 1977; Hayman and Sharp, 1981), suggests that major selective advantages to these systems are improbable at least (Mayo, 1981; cf. also Bengtsson, 1980b). Thus, chance events should be very important in the fixation of chromosomal mutations.

Given that karotypes evolve, are there trends in such evolution? In the case of the mammalian karyotype, there are three hypotheses as to the direction of change. First, Ohno (1969) has suggested that most change has been to the fusion type, with an original chromosomal set of $2n = 96$, these chromosomes being all acrocentric, with centric fusion and pericentric inversion gradually lessening the number. Todd (1967) has suggested fission as the main trend, from an initial number of $2n = 14$, with centric fission and pericentric inversion leading to many more chromosomes. Matthey (1973) has suggested that the evolution has been in both directions, i.e. increases and decreases from an original value of $2n = 40$ to 50.

Imai and Crozier (1980) reported that, overall, the mean and variance in chromosome number both rise with chromosome arm number, and that short arm size is inversely related to chromosome number, while this is not the case within particular groups. On this basis, they assessed the fission hypothesis as more likely to have been the predominant mechanism, though all known mechanisms will certainly have effected particular changes.

There is no implication in any of these models that the changes need to be adaptive. However, Bickham and Baker (1979) have claimed that karyotypic evolution is largely adaptive, on the basic assumption that "the karyotype contributes significantly to the fitness of the individual and that for a given set of biological parameters faced by an evolving lineage, there is an optimal karyotype". They have not presented evidence for the existence of such an optimum but instead have suggested what they call a canalization model for the evolution of a karyotype. This is in three stages. First, when a "lineage breaks into a new adaptive zone" there is rapid diversification of the karyotype, with poor adaptation, allowing gene and developmental regulation to be altered. In the second state, as adaptation to the new zone improves under natural selection, the process slows, and finally in the third stage, almost "all karyotypic mutations are non adaptive and the lineage becomes karyotypically stable". This third stage would be the one being evaluated in population studies of human chromosomal mutations mentioned above, the present human karyotype being at least stable, if not necessarily optimal.

Given these various models then, we can consider the evidence on the rate of change and on its adaptiveness. Imai and Crozier (1980) and Maruyama and Imai (1981) considered that in mammalian lineages there was very limited association between morphological and karyotypic evolution, but the work of Prager and Wilson (1975) does not support this conclusion for frogs and birds. Bengtsson (1980a) considered that for placental mammals only, small animals had a higher rate of karyotypic evolution than large animals, and that rich genera had a higher rate than impoverished genera. He held that reproductive biology must be important, since zygotic loss would be much more critical for animals with low fecundity than for those with high fecundity. Actual rates are given in Table 6.7 (from Gold, 1980).

Table 6.7 Evolutionary rates within genera of North American Cyprinidae

Group	Net speciation rate*	Extinction rate*	Corrected speciation rate*	Chromosome number changes/ lineage/ 10^6 years	Arm number changes/ lineage/ 10^6 years
N.A. Cyprinidae	0·40	0·25	0·65	0·012	0·056
Notropis	0·66	–	–	0·000	0·018
Other N.A. genera	0·37	0·25	0·62	0·015	0·069

From Gold (1980); cf. also Stanley (1975).
*Species per lineage per 10^6 years.

The data in Table 6.7 are on North American cyprinid fish (minnows). Gold concluded that speciation rate was not strongly associated with chromosomal number changes, which provided poor support for the punctuated equilibrium model of Eldredge and Gould (1972), i.e. the concept that evolution proceeds in fits and starts, long periods of stasis being interrupted by relatively short periods of rapid change (see Chapter 10, also Stanley, 1975, Templeton, 1980a). However, the chromosomal changes were not gradual as assessed on the basis of genetic distances. Gold concluded that the separate levels of organization seemed to be independent, or at least not directly associated. The generality of this conclusion is limited by the fact that the rates are very low compared with many other animal lineages, where changes per million years per species have been listed as ranging from 0·025 for whales to 1·395 for horses (Baker and Bickham, 1980; White, 1978; cf. also Bush, 1975; Bush *et al.*, 1977). The rate may of course be very slow indeed for certain individual chromosomes; Bickham (1981) has claimed that for eight families of Cryptodiran turtles no major change has occurred in at least two chromosomes for 200 million years, and Tegelström and Ryttman (1981) have found very limited variation in some of the large chromosomes of a wide range of bird taxa. More data on rates and their variability are needed, as was shown for amino acid substitution in Section 6.3.

6.5.1 conservation of synteny

Whatever drives evolution includes some remarkably conservative forces. We can see this very clearly in the persistence of synteny, or linkage between particular genes. Persistence of linkage may mean that the same patterns of gene-frequency change involving linked genes may be found in very distantly related taxa (cf. Hedrick, 1980).

The mammalian X-chromosome is the most remarkable known example of this persisting synteny. Ohno (1973) suggested that the evolution of dosage compensation placed strong restrictions on rearrangement between the X-chromosome and autosomes; metatherians, having a different mechanism of dosage compensation from eutherians, might be expected to have a different pattern of sex chromosome evolution. This is the case, and, as already noted, it is difficult to find great advantages for the metatherian mechanisms (Mayo, 1981).

Table 6.8 shows some X-chromosomally located genes in eutherians and one metatherian.

It is of interest that in several cases, gene order has not been preserved with synteny. In the house mouse, for example, Francke and Taggart (1980) have shown the order of two loci relative to the centromere to be centromere–*HGPRT–aGAL*, while in man the order is centromere–*aGAL–HGPRT* (cf. Pearson and Roderick *et al.*, 1979; Miller *et al.*, 1978). This change is presumably the result of an inversion.

Another remarkable case of conservation of synteny has been identified by Foster *et al.* (1981). They have identified and mapped 52 loci in the sheep blowfly *Lucilia cuprina* and have compared their linkage relationships with those in *Musca domestica* and *Drosophila melanogaster*. They conclude that "the major linkage groups have survived largely intact during the evolution of the higher Diptera". This is especially noteworthy in view of the tolerance to aneuploidy shown by, for example, *D. melanogaster*. Given a modest rate of production of chromosomal mutation and this tolerance, random change might have been expected to have been considerable.

The frequency of occurrence of new inversions in natural populations is by no means precisely delineated. Table 6.9 shows some data for a number of plant genera.

The data in Table 6.9 may be compared with the inversion frequency of 1/1730 in a wild population of the grasshopper *Moraba scurra* (White, 1961). In all these plants there were seven relevant pairs of chromosomes, so that the rate per chromosome varies from about zero in 1000 chromosomes to *ca.* 8 in 1000. In man, the rate may be considerably lower, about 0·01 in 1000 (Jacobs, 1981), but is still at least as high as the point mutation rate. This

Table 6.8 X-linked loci in a wide range of mammals.

Gene*	Pongids				Rodents		Marsupial
	Man (Homo sapiens)	Chimpanzee (Pan troglodytes)	Gorilla (Gorilla gorilla)	Orangutan (Pongo pygmaeus)	Mouse (Mus musculus)	Rat (Rattus norvegicus)	Red kangaroo (Macropus giganteus)
αGAL	+	+	+	+	+	+	+
G6PD	+	+	+	+	+	+	+
HGPRT	+	?	?	?	+	+	?
PGK	+	+	+	?	+	+	+
OTC	+	?	?	?	?	?	+

From Pearson and Roderick et al. (1979) and Cooper et al. (1979).
*αGAL αgalactosidase; G6PD glucose-6-phosphate dehydrogenase; HGPRT hypoxanthine guanine phosphoribosyl transferase; PGK phosphoglycerate kinase; OTC ornithine transcarbamylase.

Table 6.9   Frequency of inversion heterozygotes.

Genus	Number of plants surveyed	Number of inversions
	Paracentric inversions	
Aloe	123	8
Gasteria	48	2
Haworthia	72	0
Other *Aloineae*	10	0
	Pericentric inversions	
Aloe	1027	2(2)*
Gasteria	132	0
Haworthia	1046	7(2)
Other *Aloineae*	29	0

From Brandham (1977).
*Number of different inversions in brackets.

means that the supply of new inversions is unlikely to have been a limiting factor in determining loss of synteny.

Rates of change of frequency of inversions have been examined by a number of authors (Kojima, 1967; Charlesworth and Charlesworth, 1973; Bengtsson and Bodmer, 1976a). Considering the case of selection for increased frequency of an inversion in an organism where no loss of fitness arises from unbalanced gametes produced by crossing over, Charlesworth and Charlesworth showed that the probability of survival is approximately given by one-half of the square root of the loss in fitness due to recombination, i.e. the loss due to breakdown of "good" gene combinations, the so-called recombinational load. Thus, inversions might be expected to increase more slowly on the X-chromosome because of the absence of recombination in the heterogametic sex. While some organisms, such as certain species of *Drosophila*, have evidently evolved in such a manner that they can tolerate high levels of unbalanced gamete production in at least one sex so that the inversion heterozygote may be fitter than either homozygote (cf. e.g. Wright and Dobzhansky, 1946; Kojima and Tobari, 1969), this is not the case with mammals, where the fitness of unbalanced gamete producers may be as low as 0·67 (Bengtsson and Bodmer, 1976b). In this case, the analysis of Bengtsson and Bodmer (1976a) may be appropriate. Table 6.10 indicates the rate of accumulation of chromosomal mutations which are disadvantageous in the heterozygote.

It can be seen from Table 6.10 that the acquisition of an inversion in the divergence of mouse and man is most unlikely to have been purely the result of drift.

Weitkamp and Allen (1979) have shown that the synteny of vitamin D binding protein (Gc) and albumin loci has been preserved in the divergent

Table 6.10   Mean number of chromosome mutations fixed during 10^6 generations assuming a mutation rate of 10^{-5} with fertility decrease s in a population of size N.

N	s			
	0·50	0·10	0·01	0·001
10	$0·82 \times 10^{-1}$	2·7	4·7	5·0
50	$0·19 \times 10^{-10}$	$0·85 \times 10^{-4}$	3·6	4·9
100	$0·87 \times 10^{-23}$	$0·72 \times 10^{-9}$	2·5	4·7
1000	$0·2 \times 10^{-247}$	$0·17 \times 10^{-113}$	$0·80 \times 10^{-3}$	2
10 000	$0·8 \times 10^{-2496}$	$0·5 \times 10^{-443}$	$0·16 \times 10^{-41}$	$0·8 \times 10^{-3}$
1 000 000	$0·01 \times 10^{-249\ 873}$	$0·3 \times 10^{-44\ 549}$	$0·10 \times 10^{-4351}$	$0·7 \times 10^{-432}$

From Bengtsson and Bodmer (1976*a*).

evolution of three subgenera of *Equus* and *Homo sapiens*. In *Mus musculus* and *Peromyscus maniculatus*, synteny has been demonstrated for a number of genes and gene orders appear to be constant, but recombination fractions are significantly different (Snyder, 1980). This may imply that a small inversion has occurred, during the briefer time available for the divergence than that between *Equus* and *Homo*.

Evidently, while much further work is required to define rates of decay of autosomal synteny, there is already evidence that these are not constant.

TOPICS FOR DISCUSSION

1. Is it possible to relate morphological change measured from palaeontological data to biochemical change measured from comparisons of modern taxa?
2. Is is possible to remove the circularity in calculating divergence times from palaeontological data, using these to assess biochemical rates of change, and then reassessing palaeontological change thereby?
3. "[In the presence of] mutation and selection, in a large population, the mating system has no influence on the amount of genetic variability maintained by additive genes." (Lande, 1977*c*).

   Does this result have any implications for assessing the rate of change of fitness under natural selection?
4. It has been estimated that selective differences among polymorphic genotypes of isozyme structural gene loci in *Drosophila melanogaster* are of the order of $0·0005$ or less (Mukai *et al.*, 1980). Under what conditions should such selective differences be considered in investigating phenotypic change under natural selection?
5. Under what conditions is it valid to consider selective forces acting on phenotypes determined by the genotypes at a single gene locus, rather than considering all possible genotypes?
6. If linkage disequilibrium is a major factor in gene frequency change, what are the implications for phenotypic directional change?

buffered = without collisions
homoeorhetic =

7 canalization

Canalization is the term introduced by Waddington (1957) to describe one of two possible courses of evolution. In

> heterogeneous environments, varying both in space and in time, there are two major strategies in evolution. One of them is canalize, to make [the] epigenetic processes very highly buffered and homoeorhetic, with a great deal of structural stability [yielding] much the same end result whatever the circumstances . . . The opposite strategy is to have an extremely flexible development, although of course it is necessary to ensure that the flexibility produces always a rather well-adapted phenotype.
>
> (Waddington, 1969).

Wicken (1980) has suggested that the degree of homoeostasis in the "strictly thermodynamic sense", is a measure of evolutionary progress, but this argument is not soundly based (Section 2.2).

Dodson (1975, 1976) has used catastrophe theory (cf. e.g. Thom, 1975) to show that, on the assumption that "the fitness of a population is a smooth, stable function of phenotype and the environment", canalization and the potentiality for genetic assimilation (and hence quantum evolution; cf. Chapter 10) are expected to evolve. However, it is difficult to relate this to individual fitness, and therefore an explanation of canalization in individual terms is preferable.

Waddington's two "strategies" have prompted much speculation. For example, Rössler (1979) has argued that plasticity is more necessary than stability for evolution, but this is only one of the possible strategies. Plasticity has been investigated directly very little (see Grüneberg, 1980, for an example), but essentially allows change without persistence. It may be that homoeostasis and plasticity are the two faces of the same coin. For example, consider those of the widespread, insect flight polymorphisms which arise from arrested development of the flight muscles (Harrison, 1980). In some cases, e.g. Corixid bugs (Young, 1961, 1965), flying forms are the norm, but flightless forms develop at the beginning and end of the summer, the switch in development occurring at least partly through temperature changes. These flightless forms appear to be more efficient swimmers, and are only found at high frequency in permanent habitats, while where dispersal may be necessary more frequently in ephemeral habitats, flightless forms are very rare. It could be argued that the process of wing development is canalized, but with an environmentally responsive switch, or that it is plastic, in that not all insects respond similarly to the environment. (The role of the genotype is uncertain.)

The key point is that canalization comes about as a result of selection, in Waddington's view; it is not an inevitable result of physics or chemistry. And it is the kind of attribute which one would expect to be selected, since the mechanism is irrelevant, i.e. any mechanism which will achieve the adaptation will be satisfactory. Thus, compensation for different numbers of X-chromosomes works differently in eutherian and metatherian mammals, and differently again in Diptera, in each case constituting a mechanism of dosage compensation. On the other hand, the evolution of structures must occur within whatever framework exists. Thus, three cranial nerves serve the motion of the eyes in man, one serving four muscles, and two serving only one muscle each. As the argument of Chapter 5 should have made clear, there is no reason to suppose that this division of labour is adaptively optimal. (This is of course a well-recognized fact; see e.g. Gould (1980), or Darwin (1859, Chapter 14):

> why should the brain be enclosed in a box composed of such numerous and such extraordinarily shaped pieces of bone, apparently representing vertebrae? As Owen has remarked, the benefit derived from the yielding of the separate pieces in the act of parturition by mammals, will by no means explain the same construction in the skulls of birds and reptiles.)

A problem immediately arises, in how to measure the degree of canalization. On the hypothesis (advanced for other reasons) that heterozygotes are better canalized than homozygotes, one might expect that heterozygotes would be less variable than homozygotes for a range of quantitative traits. There is some evidence for this hypothesis (Mitton, 1978; Eanes, 1978; Mayo *et al.*, 1980) and some against it (Handford, 1980).

Another approach is to use asymmetry as a measure of homoeostasis (Thoday, 1958). Again, the evidence, though in accord with the hypothesis, is less than compelling (Jantz and Webb, 1980).

We now consider two examples in more detail, the first relating to canalization through process, the second to canalization of a trait.

7.1 DOSAGE COMPENSATION AND DUPLICATE GENE INACTIVATION

When an amphiploid is formed in plants, the homoeologous structural genes are not inactivated very rapidly. In common wheat, *Triticum aestivum*, Hart and Langston (1977) identified at least eight, perhaps ten, triplicate sets of homoeologous genes and only one structural gene which apparently had no homoeologues, but which might in fact have been a diverged duplicate on one of the triplicate sets. Thus, given that of 33 genes, 22 at most could be inactivated, between 0/22 and 2/22 have been. This is a rate of between 0 and 10^{-5} inactivations per generation, assuming wheat to have been formed approximately 10 000 years ago (Feldman, 1976). As Hart and Langston pointed out, this rate is not inconsistent with the view that multiple loci allow inactivation through the accumulation of genes which would otherwise be lethal (Muller, 1914; Nei, 1969; Ohta and Kimura, 1971).

In tetraploid fish, Bailey *et al.* (1978) have suggested that 50% of duplicate loci have been silenced in 50 million years. Allendorf (1979) has taken a slightly higher rate as more probable, and has analysed a model of selection favouring an intermediate optimum in enzyme activity. This would yield a much faster rate of loss of activity in populations of moderate size, i.e. $N \leqslant$ 2000. With regard to the suggested average generation times, and other factors, these analyses imply a rate of gene silencing at least of an order of magnitude slower than in hexaploid wheat. Bailey *et al.* (1978) suggested that the probability of fixation would be approximately one-half after $15N + \mu^{-3/4}$ generations, which would imply a very large value for N, i.e. $> 10^5$.

The development of highly heteromorphic sex chromosomes in many of the relevant species of fish appears to have been very slow, implying that dosage compensation too may not be highly evolved, but none the less the apparent difference in rates is striking. It is not clear that hexaploidy as against tetraploidy is the explanation. Certainly what has been established about the population genetics of higher order polyploidy does not suggest this.

In the case of duplication through misreplication, there is evidence from the families of duplicate genes studied in man that activity of several loci has persisted, even to the extent of positive maintenance of sequence homology in the $\alpha 1$ and $\alpha 2$ loci of human α-globin (Liebhaber *et al.*, 1981).

In general, of course, as has long been recognized (Haldane, 1932; Bridges, 1935; Stephens, 1951), duplication provides the largest single source of raw material for evolution. The duplicated sequence can evolve outside the constraints imposed by the function of the gene product of the original sequence; evolution can change rate as well as direction (Ohno, 1970). In addition, duplication provides the possibility of duplicated regulatory mechanisms, thereby avoiding some of the problems in the regulation of novel functions which have been highlighted by Clarke (1980) and Hall (1980). Thus, some apparent gene silencing may represent divergence of function to such an extent that the gene products are not recognized as having originated in duplicate sequences. Britten and Davidson (1969) suggested that eukaryotic regulation might proceed by integrator genes being transcribed into activator messenger RNA, followed by activation of structural genes through binding of the activator mRNA to acceptor sites adjacent to structural genes, with large numbers of such integrator sets in the genome. If a model of this kind is correct, and such models seem more difficult to test than to construct (e.g. see Nebert, 1981, for an application of this approach to elucidating the huge range of types of cytochrome P-450 of importance in pharmacogenetics), then duplication has been an essential element in the evolution of regulation.

7.2 CLUTCH SIZE AS A CANALIZED TRAIT

In developing his theory of the determination of clutch size by natural selection, Lack (1947) proposed that the mean clutch size observed for any given species in any given environment corresponded to the largest mean number of young for which the parents could provide food. While this theory has been strongly criticized, and to some extent modified, it is now generally accepted (see Owen, 1977, for a review).

However, it is notable that, with few exceptions, variability in clutch size is rarely discussed or treated in considering this important evolutionary problem (Cody, 1966). For example, Charnov and Krebs (1974) noted specifically that "Lack's hypothesis made no prediction about variations round the most productive clutch", but did not incorporate variance into their theory of optimal clutch size, defined as that which maximizes an individual's fitness. Where another type of variability, clutch overlap, has been examined, constant clutch size has been used as a limitation on experimental variability (Burley, 1980).

Similarly, very few investigations of the inheritance of clutch size have been conducted, though this is obviously of considerable importance in assessing the theory. However, a bird with very variable clutch size (Perrins and Jones, 1974), the great tit, *Parus major*, had a narrow sense heritability of clutch size of approximately 0·5, i.e. the proportion of total variance in clutch size attributable to the additive effects of genes was about 50%, a very high value. It would be of great interest to have similar values for birds with highly invariant clutch size, but in such cases very many data would be needed.

Gillespie (1974) has considered a model for the genetic determination of both location and dispersion of offspring number, which in fact allows only very tightly constrained fecundity distributions to be stable (Mayo *et al.*, 1978). Apparently, in general, an increase in the variance in offspring number of a genotype decreases its fitness (Mayo *et al.*, 1978; Gillespie, 1977).

The model of greatest interest is that of Heyde (1978) and Heyde and Schuh (1978). Unlike Gillespie, Heyde has used an infinite population size model, but he has shown that maximization of the probability of survival of the species to a particular time, t, can occur through the evolution of a characteristic clutch size fixed at either a single value, k, or a pair of values, $k, k + 1$.

Heyde's model requires that there be a constant "maximum number of young which the parents can raise on the average". If there is any variation about this value, it follows that in some generations more young than this mean number must be raised. Accordingly, it must be some other property of clutch size, such as the actual physical or physiological upper limit on the number of eggs which can be laid, which is the ultimate determinant of clutch size distribution. This has been suggested in a number of cases (Wagner, 1957; Haukioja, 1970).

None of the models discussed above covers the case of the shore birds shown in Table 7.1

The simplest description of the distributions shown in Table 7.1 is to say they are canalized, though not necessarily at the same value (cf. *Charadrius alexandrinus* in Australia and North America). The question of interest is what kind of process yields these triangular or near-binomial distributions.

Table 7.1 Clutch sizes recorded in RAOU and NANRS records.

	Species	Clutch size					
		1	2	3	4	5	6
Australia	Banded plover *Vanellus tricolor* *(Zonifer tricolor)*	5	5	27	86	0	0
	Spur winged plover *Lobibyx novae-hollandiae*	21	47	117	300	2	1
	Red-capped dotterel *Charadrius alexandrinus*	26	168	3	0	0	0
	Black-fronted dotterel *Charadrius melanops*	8	38	62	1	0	0
USA	Snowy plover *Charadrius alexandrinus*	3	5	11	1	0	0
	Piping plover *Charadrius melodus*	1	4	7	30	0	0
	Wilson's plover *Charadrius wilsonia*	6	4	19	0	0	0
Scotland	Dotterel *Eudromias morinellus*	2	33	376	1	0	0

From Mayo (1980) and Nethersole-Thompson (1973).

McLean (1972) has suggested that four was the ancestral clutch size of the Charadrii, as no member species "normally lays a clutch of more than 4, regardless of its degree of specialization". It is not clear (cf. Table 7.1) that this is less rigid than, say, in albatrosses (Tickell and Pinder, 1966). However, if we ignore the small numbers of clutches that are larger than the modal size, we may regard deviations downwards within a species as the raw material on which natural selection has acted in speciation.

Treating clutch sizes of 4, 3, 2 and 1 as deviations of 0, 1, 2 or 3 from the mode, we may examine the fit to one process which might describe, if not explain the variation. This is the "burnt fingers" process, a name applied by Arbous and Kerrich (1951) to the case of accidents in industry, where, until an accident occurs, an individual has a given, constant chance of having an accident, and, when this happens, immediately "learns" enough to lower his chance of having further accidents. This crude model allowed the development of a satisfactory distribution where, in time $(0, t)$, the probability of the first accident occurring is δ, but thereafter falls to ε, i.e. $1 > \delta > \varepsilon > 0$.

These parameters may then be estimated, and I have done this for the clutch size of the banded plover, spur-winged plover and piping plover, with results which conclusively are not in agreement with the model (Mayo, 1980). An implication of the model might be that if the clutch sizes of particular species were simply the result of random fixation of genetic variation lowering clutch size, the distribution of modal clutch size among species would also be approximately triangular. Table 7.2, condensed from Table 1 of McLean (1972), shows that this is not the case.

Table 7.2 Mode or range of clutch size of 319 species of Charadrii

Modal clutch size	Number of species
4–6	1
4–5	1
4	137
3–5	3
3–4	16
3	25
2–4	19
2–3	32
2	64
1–4	3
1–3	4
1–2	9
1	5

From McLean (1972).

Little, if anything, is known of the relationship of the cost of reproduction to variability in reproductive performance,. yet this cost is critical (Fisher, 1930*b*; Calow, 1979). Until more is known in this area, very general models, like that of Heyde mentioned above or that of Grey (1980) in which clutch size variability is supposed to minimize the probability of population extinction but which yields very inconclusive results, are unlikely to be helpful in assessing the origin of clutch size canalization.

7.3 IMPLICATIONS OF CANALIZATION

7.3.1 directional selection

The work of Rendel's school shows that genetical variability may persist, concealed by canalization, and unavailable for direct selection (cf. Rendel, 1965). Mutants which allow this variability to be expressed allow directional selection to move the location of the canalized trait outside its previous bounds.

For example, the scutellum of *Drosophila melanogaster* normally bears four bristles. In scute mutants, the number is reduced, quite often to zero, and there is substantial variability about a mean which varies between stocks. In some cases, flies with more than four bristles are not rare. These stocks respond to selection for increased bristle number and the mean is raised rapidly (Rendel and Sheldon, 1960; Rendel, 1965).

The variability hidden by canalization may also be revealed by environmental stresses, as in the "genetic assimilation" work of Waddington (1953). Here, heat shock or ether shock to larvae was used to produce particular rare phenotypes and flies showing the rare phenotype after stress were then used to produce the next generation. After he had done this for a number of generations, breeding each time from the flies which responded to the environmental stimulus, Waddington found that of animals reared without the larval stress, a substantial proportion manifested the trait.

Thus, considerable variability may persist in a population, providing the potentiality to respond to sudden, marked, unusual stresses, with the possibility of rapid evolution thereafter.

7.3.2 random genetic changes

Since canalization can mask genetical variability, it is to be expected that nearly neutral mutants could rise to quite substantial frequencies by chance, without affecting the normal range of the phenotype. This would mean that individual responses to large variations in the environment of the kind mentioned in Section 7.3.1 would depend, in part, on cryptic variability which previously had not been adaptively relevant.

Dosage compensation has already been discussed; this could allow not just nearly neutral mutants to rise to substantial frequencies, but lethals and highly deleterious mutants. The well-established properties of polysomic inheritance (cf. Bennett, 1976; Mayo, 1971) also include this factor; i.e. the gene frequencies at which mutation and selection balance each other are far higher for polysomic inheritance than for diploids.

7.3.3 stabilizing and normalizing selection

Waddington (1953) suggested that canalization might arise through stabilizing selection, but such selection is now regarded as leading to loss of variability (Robertson, 1956) though not necessarily at a high rate (Mayo and Hancock, 1981). Thus, it is unlikely that canalization has arisen as a result of stabilizing selection, except in so far as this would imply directional selection for invariant response (Ho and Saunders, 1979). Ho and Saunders have further claimed that the invariant response would have been transmitted cytoplasmically, rather than determined by nuclear genes. It is not clear why this should be so (cf. Section 7.3.4).

Once canalization is well-developed, stabilizing and normalizing selection may not be expected to have much influence on canalization itself, though they are in fact still significant, as was shown in Chapter 5. These types of selection are therefore largely irrelevant to the maintenance of canalization apart from selection against extreme (decanalized) deviants, which may in any case be disadvantageous as such, or through pleiotropy (e.g. the extra digits frequently found in the polycystic kidney syndrome in humans).

7.3.4 selection for dominance

When Fisher (1928a, b) introduced his theory of the evolution of dominance through modification of the response of the wild type to recurrent mutations, he considered the case of modification in the heterozygote of particular individual genes but there was no implication that the modifiers would be completely specific to each gene in every case. Later, as a result of the criticisms of Wright (1929), Haldane (1930) and others, the evolution of dominance came to be discussed almost completely in terms of the modification of each particular gene by a specific set of modifiers. Haldane's (1930) alternative model, selection among differing alleles with the most active thereby becoming the most frequent through its "factor of safety" in the heterozygote, was also ignored, though as recent work of Cavener and Clegg (1981) shows, it may occasionally be important. As Sved and Mayo (1970) pointed out, it might be that what actually evolved was an overall set of diploid responses, covering dosage compensation, dominance and substantial differences in activity of the products of most structural genes. The failure to detect major phenotypic effects of isozymes differing considerably in activity as against those not so differing in humans (Beckman, 1978) would tend to support this view.

Kacser and Burns (1981) have presented the case for dominance and recessiveness, especially the recessiveness of inactive mutants, being the "inevitable consequences of the kinetic structure of enzyme networks". In particular, a heterozygote having only 50% activity of an enzyme would not be expected to be phenotypically detectable. However, this theory has no implications for dominance breakdown when crossing populations for traits known to have been selected strongly (see Kettlewell, 1973, and Ford, 1975, for references); this is predicted on the modification theory. Nor does it predict that a known dominant phenotype might be the result of the combination of several recessives (Stern, 1929; Fisher, 1931). Accordingly, Fisher's hypothesis may still be relevant for predictive purposes in particular cases, e.g. all known cases of mimicry in the Lepidoptera.

It is not easy to calculate, in a simple way, the effect of selection for a general buffering genotype (Sved and Mayo, 1970). For small selective forces, however, one can readily show that they will be approximately additive, i.e. the conditions of gene frequency equilibrium will be met sufficiently to yield reasonable approximations.

To take the simplest case, consider the three-locus analogue of Wright's example of Fisher's model. Initially, suppose that selection acts on two independent loci, rather than one, and that a third, intrinsically neutral locus produces dominance (for fitness) at both the other loci, as shown in the following table:

	BB			Bb		
	AA	Aa	aa	AA	Aa	aa
MM	1	1	$1-s$	1	1	$1-s$
Mm	1	1	$1-s$	1	1	$1-s$
mm	1	$1-gs$	$1-s$	$1-ht$	$(1-gs)(1-ht)$	$(1-s)(1-ht)$

	bb		
	AA	Aa	aa
MM	$1-t$	$1-t$	$(1-s)(1-t)$
Mm	$1-t$	$1-t$	$(1-s)(1-t)$
mm	$1-t$	$(1-t)(1-gs)$	$(1-s)(1-t)$

Ignoring the M locus, the initial frequencies of the alleles a and b at the two loci are approximately

$$q_B = \frac{\mu_B}{ht}$$

$$q_A = \frac{\mu_A}{gs},$$

where μ_A and μ_B are the rates of mutation from A to a and from B to b respectively. Assume for simplicity that there is equilibrium for gene and genotype frequencies at birth. We want to calculate Δq_M. This is given by $q_M' - q_M$, where

$$q_M' = \frac{\varphi q_M + \theta q_M^2}{\varphi + \theta q_M^2}$$

$$\varphi = (1 - sq_A^2)(1 - tq_B^2)$$

$$\theta = -2gsp_Aq_A(1 - 2htp_Bq_B - tq_B^2) - 2htp_Bq_B(1 - 2gsp_Aq_A - sq_A^2).$$

Then

$$\Delta q_M = \frac{\theta q_M^2(1 - q_M)}{\varphi + \theta q_M^2}.$$

Similarly, we can derive Δq_A and Δq_B. For example,

$$\begin{aligned}
\Delta q_A = &(((1 - s)q_A^2(1 - 2htp_Bq_Bq_M^2 - tq_B^2) \\
&+ p_Aq_A(1 - gsq_M^2 - (ht - ghst)2p_Bq_Bq_M^2 + gstq_B^2q_M^2)) \\
&- q_A(\varphi + \theta q_M^2))/(\varphi + \theta q_M^2).
\end{aligned}$$

Now we can suppose, by analogy with the single locus case, that $\Delta q_A = 0$ and $\Delta q_B = 0$, and we can solve these equations for q_A and q_B. If, for simplicity, we suppose

$$\mu_A = \mu_B = \mu$$
$$g = h$$
$$s = t$$

then

$$\begin{aligned}
&q_A^2(-2 - 2h + 2hs + 8hq_M^2 + 4h^2s^2q_M^2) \\
&+ q_A(-1 - 5hq_M^2 + 2h^2sq_M^2) - hq_M^2 = 0.
\end{aligned}$$

Simplifying, let

$$q_A = \frac{\mu_A}{gsq_M^2}$$

and

$$q_B = \frac{\mu_B}{htq_M^2}.$$

Then, substituting in the expression above for Δq_M, we have, approximately,

$$\Delta q_M = -2(1 - q_M)(\mu_A + \mu_B),$$

which is analogous to the single locus formula known not to be seriously inaccurate for values of q_M down to about $0 \cdot 4$.

A similar result holds for three loci having their phenotypes modified by a fourth. Thus, it seems that (as one might expect) the small selective forces acting to produce dominance will be approximately additive. A general buffering genotype, or a set of genes acting as "enhancers" in this manner, might thus be at an advantage somewhat greater in magnitude than $0 (\mu)$.

TOPICS FOR DISCUSSION

1. Is it possible to identify canalizable traits solely on the basis of what traits are canalized? That is, can we distinguish between traits whose development is canalized and others for which the appearance of canalization results from other factors, e.g. their discrete nature?
2. How can one distinguish between Ohno's "frozen accident" hypothesis of X-chromosomal gene content (Chapter 6) and the hypothesis that the X-chromosomal gene content reflects a significant adaptation?
3. "Should there be a positive or negative correlation between survival of adults in a bird population and their clutch size?" (Högstedt, 1981).
4. The cowbird, *Molothrus ater*, is a brood parasite which lays its eggs in the nests of over 100 other species. This "has led to certain assumptions about the role of species-typical stimulation in its development. It has often been proposed, for example, that cowbirds may lack sensitivity to certain forms of early experience, thus reducing the risk of incorrect mate identification." (West *et al.*, 1981).

   Suggest a testable genetical hypothesis for the evolution of such insensitivity.

8 processes regarded as distinct from natural selection

Darwin recognized that natural selection was not the only possible determinant of evolutionary change. For example, he pointed out that some variation might be selectively neither advantageous nor disadvantageous, a fact recently emphasized by Freire-Maia and Robson (1979). As noted in Chapter 6, such neutral phenotypes may drift apart in frequency and contribute very significantly to population differentiation. More importantly, perhaps, Darwin (1859, 1871) saw sexual selection as distinct from the process of the "preservation of favoured races in the struggle for existence". More recently, selection among genetically related groups and selection favouring senescence have been suggested to be processes distinct from natural selection. These suggestions will be examined critically here.

8.1 SEXUAL SELECTION

Single locus, gametophytically determined self-incompatibility gives almost complete protection against self-fertilization with limited protection against less close inbreeding and a high probability of reproductive success in small populations of plants. The existence of two-, three- and even four-locus systems is therefore difficult to explain (Mayo, 1978). It may be that at some stage in the evolutionary history of the plants possessing the more complex systems, some reproductive advantage was gained for plants having duplications engendering the novel system.

Such examples apart, there is one type of selection which is confined to animals. This is sexual selection. Darwin (1859) defined the process as follows: "[Sexual selection] depends, not on the struggle for existence, but on a struggle between males for possession of the females; the result is not death to the unsuccessful competitor, but few or no offspring." Darwin (1859, 1871), and later Fisher (1930b), used the theory to explain many cases of sexual dimorphism. Darwin argued that the selection occurred largely by the more successful males obtaining the females, but Fisher added the very important concept that female preference might allow evolution in males. In particular, female preference might allow the evolution of apparently use-less but (to human eyes) highly decorative organs, patterns or displays.

The qualitative arguments of Darwin and Fisher have now been sup-ported by extensive theoretical analysis by O'Donald (summarized by O'Donald, 1980), while many of the predictions of the theory have to some extent been borne out by experiment (cf. O'Donald, 1980), despite the difficulties associated with detecting and measuring the strength of sexual selection (Nadeau *et al.*, 1981). One of Fisher's arguments remains of particular interest. He suggested that as female mating preferences evolved jointly with the secondary sexual characters of males, the process could become self-reinforcing and accelerate until checked by strong selection against the more extreme forms of the male traits. O'Donald showed that this would occur, using a two-locus model, one for trait variation in the male, one for mating preference in the female. Lande (1981) has extended this to the much more realistic case of polygenically determined variation in both sexes, and has confirmed Fisher's prediction. While several aspects of Lande's work await complete elucidation, sexual selection, although still selection, is a very convincing example of apparently directed evolutionary change, at an accelerating rate, determined by a well understood and unexceptional genetical mechanism.

8.2 KIN SELECTION

Individual selection, the topic of the preceding four chapters, is the process whereby a relative advantage to an individual in some aspects of viability or fertility allows it to leave more offspring than the average. Obviously, this is not only a relative advantage but also involves an interaction between the individual and its fellows. The key point, however, is that the advantage accrues to the individual, and the interaction is not assessed otherwise. Consider now the case where manifestations of a trait are such that ad-vantage, if any, accrues not to the individual but to some group to which it belongs. Does natural selection apply to such traits in the same way as before? For Fisher (1914) it did:

> From the moment that we grasp, firmly and completely, Darwin's theory of evolution, we begin to realize that we have obtained not merely a description of the past, or an explanation of the present, but a veritable key of the future; and this consideration becomes the more forcibly impressed upon us the more thoroughly we apply the doctrine; the more clearly we see that not only the organisation and structure of the body, and the cruder physical impulses, but that the whole constitution of our ethical and aesthetic nature, all the refine-ments of beauty, all the delicacy of our sense of beauty, our moral instincts of obedience and compassion, pity or indignation, our moments of religious or

mystical penetration – all have their biological significance, all (from the biological point of view) exist in virtue of their biological significance.

A very sweeping claim, and not one which Fisher himself justified to any great extent. However, more recently the concept of kin selection has been introduced to account for the evolution of traits such as obedience or compassion. These are traits based in the kind of interaction which, as noted above, is ignored in usual discussions of fitness (cf. Sections 6.1 and 6.2). It can be included by using what is termed inclusive fitness (cf. Hamilton, 1972, for discussion). The inclusive fitness effect of a genotype A is the sum of all the effects of that genotype on the individual bearing A and all the other individuals in the population, the effects being weighted in every case by the relatedness between A and the other. Inclusive fitness will, in most cases, be even less susceptible of measurement than individual fitness, so that kin selection is approached in other, less direct ways.

Maynard Smith (1964) introduced this concept as a substitute for group selection. (It was also implicit in the work of Hamilton (1963).) Group selection is selection among distinct, isolated breeding groups, whereby groups containing individuals who act for the good of the group are at an advantage to other groups where this is not the case. Kin selection, by contrast, requires that individuals within a population behave differently to other individuals according to how closely related they are to those individuals. It requires further that individual organisms be able to recognize their relations and to distinguish degrees of relatedness. Evidence for the first of these two requirements is accumulating. For example, Blaustein and O'Hara (1981) showed that Cascades frog (*Rana cascadae*) tadpoles associate preferentially with siblings whether or not they have been reared together. These results confirm those obtained in another amphibian and two species of mammals. Thus, recognition of relatedness appears not to be dependent necessarily on early learning through propinquity.

As an example of kin selection, Hamilton (1964) showed that a gene which conferred a disadvantage of magnitude c to its possessor but an advantage b to some other individual would increase in frequency if $c/b < r$, where r was a measure of the degree of genetical relatedness of the two individuals. Griffing (1981a, b, c, d) has generalized this result to $r(n-1) > c/b$ for interacting groups of size n, rather than 2. Wade (1980b) has shown that the change in gene frequency under kin selection is the total of two components, one resulting from individual selection, the other resulting from group selection, the groups here not being reproductively isolated. For altruistic traits increased in frequency by kin selection, the component relating to individual selection is always negative and the other is always positive. Michod (1979) showed that inbreeding alters the simplicity of Hamilton's rule, to the extent that a stable polymorphism for behaviour may be possible, generally when the two individuals involved in the altruistic act are both substantially inbred, at least under weak selection.

The model considered by Hamilton and others was that of a trait determined by alleles of a single gene, a convenient analytical device but not necessarily relevant to most behavioural traits. However, Yokoyama and Felsenstein (1978) have extended the analysis to the case of a quantitative trait and have established the validity of a rule similar to Hamilton's. If environmental variability in the trait also occurs, as may be expected, then the situation may change. Boyd and Richerson (1980), for example, have shown that if selection on the trait is strong and heritability is very low, the mean level of the trait may be higher than that predicted by the analogue of Hamilton's rule. This resembles the case of family selection in animal breeding, where selection is on the basis of family rather than individual performance for traits of low heritability (Falconer, 1960). (This is because the family mean is a better predictor for breeding purposes than the individual mean where variability is mostly environmental.)

Wade (1980a) and others have shown experimentally that kin selection can indeed be important in influencing behaviour. However, both the breeding structure of a population and the pattern of interaction between individuals influence the evolution of behaviour, so that group and kin selection are not as distinct as Maynard Smith implied.

Given the evidence of behaviour patterns which benefit a genetically related group at the expense of the individual, it may be expected that individuals not manifesting the behaviour but benefitting from it in others will increase in frequency (cf. e.g. Haldane, 1932; Alexander, 1974). However, Wade and Breden (1980) have shown that for altruism to exist

$$\frac{(c-b)}{n} + \frac{b}{2k} > 0,$$

(n being group size divided into k families) whereas for contrary behaviour to increase, the sign of the inequality must be reversed. Thus, in a population where altruistic behaviour has already evolved, contrary behaviour is unlikely to arise. More generally, Axelrod and Hamilton (1981) have shown that co-operation can evolve in an asocial environment, and similarly resist the increase of contrary behaviour.

In any science, facts accrue like interest on a bank deposit or barnacles on a jetty's piles, and are then subsumed in a theory which sweeps away the need for so much detail. In biology, however, the models are often so simple compared with the reality that they require amendment before they are fully established. This is true of kin selection.

From the theoretical point of view, Cavalli-Sforza and Feldman (1978) have argued that the inclusive fitness approach is approximate, and as such is inferior to customary population genetical modelling and likely to yield misleading results. While the differences are disputable (Maynard Smith, 1980) and in any case modest (Feldman and Cavalli-Sforza, 1981), there is evidently a need to reconcile these approaches.

In terms of testing the theory experimentally, Lester and Selander (1981) states that "Hamilton's theory predicts that the occurrence of altruistic behaviour should correlate with the likelihood that the recipient is closely related to the actor". They therefore attempted to determine whether colonies of the social insect *Polistes* were so structured that workers were highly related to queens to compensate for their inability to reproduce. Their consequent null hypothesis was $r > \frac{1}{2}$, since sisters are more closely related than mother and daughter and hence females should rear sisters in preference to daughters. However, in many cases r was significantly less than $\frac{1}{2}$, whereas in other eusocial insects it was greater than $\frac{1}{2}$. The elaborations of Hamilton's theory discussed above do not predict these results, but (assuming that they are validated in further experiments) what changes are necessary is not clear. However, wholesale rejection of the theory, as advocated by Darlington (1981), hardly seems justified. Many of Darlington's criticisms had, in fact, been answered by Dawkins (1979), though it is still possible to regard kin selection as merely a special case of natural selection, advantages accruing indirectly to individuals, rather than directly. To do so is to obscure rather than clarify, so that it is preferable to distinguish the two possibilities.

8.3 EVOLUTION OF SENESCENCE

Many organisms exist which do not age; they are all single-celled organisms, but some single-celled organisms do age. Virtually all multi-cellular organisms age. Thus, it is evident that senescence, i.e. the process whereby the probability of death increases with increasing age, has evolved, or that it is a primitive property of an organism which has been lost only from some single-celled organisms, which seems unlikely. Certainly there is genetical variation in many components of senescence (Lints, 1978). Evolution of senescence could have come about in one of two ways: either senescence could itself be adaptive to some extent, or it could have evolved as a consequence of the evolution of other characters. If ageing has evolved as an adaptive trait, then it has presumably done so because the advantages of short generation interval and limited competition by post-reproductive individuals for resources have been greater than the disadvantages associated with increasing numbers of debilitated individuals in the population. Guthrie

(1969) has reviewed the evidence for the adaptiveness of senescence in animals. He has suggested that it is broadly the result of stabilizing selection, for relatively constant population structure. Overall, one may conclude that there is some support for this as a mechanism of evolution of senescence, but that it is not a complete explanation.

It is certainly not appropriate to plants (Thimann, 1980). Senescence of part of plants, i.e. seasonal senescence, which extends to the whole organism in the case of annuals, is certainly an adaptation, whatever its origin, to the climatic changes which we call seasons. Furthermore, annuals and perennials are both adapted to the annual seasonal cycle despite their quite different life cycles. Because of the near universality of plant senescence, one need not seek extraneous explanations for its existence.

Haldane (1940) was perhaps the first to consider senescence in animals as a by-product of evolution through natural selection for other traits. He was considering the variation in age of onset of human disorders, and held that "differences in age of onset in man probably correspond with differences in penetrance in *Drosophila* or other insects". Examining the inheritance of age of onset in a number of families segregating for different disorders, he concluded that "modifiers are presumably being selected which delay the age of onset" in Huntington's chorea. More generally, he concluded that selection for delayed age of onset would be a similar process to the selection for dominance which Fisher had suggested accounted for the evolution of dominance. Thus, for any particular gene it would be a second order process, but overall might be expected to be selected more strongly as part of selection for general buffering (see Chapter 7). Evidence is wholly lacking, but further investigation of ages of onset of different disorders in different populations might allow Haldane's hypothesis to be tested.

A development of this idea is associated with Williams' work (1957), who suggested that senescence might have evolved by selection of genes having different effects on fitness at different ages. Medawar (1952) had essentially defined this theory in some detail:

> if hereditary factors achieve their overt expression at some intermediate age of life; if the age of overt expression is variable; and if these variations are themselves inheritable; then natural selection will so act as to enforce the postponement of the age of the expression of those factors that are unfavourable, and, correspondingly to expedite the effects of those that are favourable.

This has been tested theoretically, and it has been shown that an early advantage (in viability or in accelerated sexual maturity) can be balanced by a late disadvantage (i.e. early senescence), but that the balance is less stable than that of a classical, balanced polymorphism having the same total selection intensity (Mayo, 1973). Hirsch (1980) has also shown that changes in the probability of death with age may be sufficient to increase the mean population fitness; this will occur in a population which is increasing in number, but may not occur in a population of constant or decreasing size.

Rose and Charlesworth (1980) have suggested that Medawar's theory rests primarily on the concept that "deleterious mutations exerting their effects only late in life would tend to accumulate, because of their minimal effects on fitness". They contrast this with the pleiotropy theory, attributing to Williams (1957) the view that "many of the genes with beneficial effects on early fitness components have pleiotropic deleterious effect on late fitness components, but are nevertheless favoured by natural selection." As we have seen, the mutation accumulation theory is a variant of Haldane's theory, while Williams's is a variant of Medawar's. However, it is of some interest that Rose and Charlesworth (1980) found that selection experiments with *Drosophila* and *Tribolium* generally support the pleiotropy theory, and do not support the mutation accumulation theory (cf. also Rose and Charlesworth, 1981*a*, *b*).

Whatever the complete picture may be, we are evidently seeing here the operation of powerful limits on the process of evolution by natural selection. Other things being equal, lengthening of the generation interval must slow the process of evolution by natural selection, while several different processes which are modified by natural selection, e.g. early viability, time to sexual maturity, clutch size, are such that improvements to them may be associated with later debility or reproductive inadequacy. As Rose and Charlesworth showed, increases in reproductive output at late ages are associated with decreases in early output, diminished overall reproductive rate, and increased longevity.

TOPICS FOR DISCUSSION

1. Does kin selection lead, in general, to regular mild inbreeding, and if not, why not?
2. In attempting to assess the role, if any, of natural selection in the evolution of senescence, what differences should be sought between animals which age, such as mammals, and apparently non-ageing animals, such as fish?
3. "Cooperation and competition are generally viewed as being virtually opposite extremes. My results suggest the reverse. In certain cases, interspecific competition may provide the very selective pressures that lead to the evolution of [intraspecific] cooperation." (Buss, 1981).
    Develop a genetical model for the basis of such co-operation and suggest how it might be tested.

9   sexual reproduction

9.1 DIRECT CONSEQUENCES IN RELATION TO SELECTION

Sexual reproduction is not necessary for evolution by natural selection, but is has certain consequences for the way in which such selection can occur. Furthermore, given the existence of sexual reproduction, species can be defined as sexually reproducing populations reproductively isolated from each other, and speciation then becomes a process barely distinguishable from the development of reproductive isolation. This is not a problem with asexual organisms, since here every individual is in principle taxonomically distinct.

Sexual reproduction has the consequences of segregation and recombination. These processes limit the possibility of accumulation of balanced linkage combinations. Such combinations can arise either through a system of balanced translocations as in *Oenothera lamarckiana*, through duplication, or through close linkage.

The system of balanced translocations is, at least superficially, not a state from which evolutionary progress is likely to be great. Duplication, though very important, will be disadvantageous in many cases through the unbalanced gametes and zygotes which arise. Closer linkage, however, does not have inherent disadvantages of these kinds. Accordingly, one might argue that very tight linkage should be the rule. Fisher (1930*b*) and Muller (1932) both advanced arguments why recombination should be advantageous, the main reason being that, in a population having recombination, its existence will allow the fixation of mutants of different genes to be to some extent independent. However, if there is no recombination, two new mutants can only be fixed if one arises in a descendant of the other. Thus, evolution will be much slower. Maynard Smith (1971) suggested the additional but related advantage that mixing of different populations on colonization of a new environment would allow much more successful adaptation to the new environment. It has also been argued that recombination will be most advantageous in marginal populations of a single species (Williams, 1957; Layzer, 1980; Case and Bender, 1981), which could mean that elevated frequencies of recombination would be maintained because of the higher rates of speciation in marginal populations. Chapman (1981) has presented evidence, from a comparison of synchronously and sequentially hermaphrodite sea-bass (Serranidae) taxa, that there is no direct association between the degree of recombination and the rate of speciation. He argued that "synchronous hermaphrodites have a higher potential rate of recombination than sequential species", essentially because of the higher effective population size of the former. He then showed for several sea-bass genera that this was not the case; rather there was an association between "recombination potential" and variance in speciation rate. This is in itself a very valuable result (cf. Section 6.4), though not conclusive because factors other than effective population size may be critical and because the definition of recombination potential is not completely unambiguous.

Felsenstein (1974) and Felsenstein and Yokoyama (1976) have extended and confirmed the early arguments of Fisher and Muller; in theory, recombination can indeed be adaptive, though selection for closer linkage may be more likely. One curious result is that in most cases, recessives which increase recombination are expected to be favoured, rather than dominants. Evolution of tighter linkage was first suggested by Fisher (1933a) for a specific case of strong artificial selection; in general, he regarded the tightening of linkage as the more probable direction of change. Conditions under which this might occur naturally are very complex, depending on the individual properties of the major loci which are linked, on their interactions and on the degree of disequilibrium between them (Feldman et al., 1980).

A second cost of sexual reproduction lies in the unbalanced nature of the contributions of the two sexes, where males, in many cases, contribute little more than their genes. Specialized castes in social insects, specialized sexual roles and widely dimorphic sexes are phenomena best explained on the basis of adjustment of parental investment in reproduction towards an optimal return (Fisher, 1930b), though sexual selection (Darwin, 1859; Fisher, 1930b) and kin selection (Hamilton, 1967; Colwell, 1981) may also be involved. Major random contributions to such differentiation have not been suggested and would be hard to explain.

9.2 REPRODUCTIVE ISOLATION

Given the wide acceptance of Mayr's (1969) definition of species, that is "species are groups of interbreeding natural populations that are reproductively isolated from other such groups", reproductive isolation is evidently the prime criterion for speciation. Accordingly, we shall consider Dobzhansky's (1970) classification of types of reproductive isolation in order to assess the importance of natural selection in each category and overall.

Before doing so, however, we will consider the relative importance of rapid evolution of reproductive isolation with geographical isolation and without it. Geographically isolated populations subjected to different selective pressures may change reproductive traits through pleiotropy or sexual selection. They may also change by chance. Indeed, it is virtually certain that they will change, both from selection and from chance effects (Campbell, 1981; Griffiths, 1981; Kimura and Weiss, 1964). Unless the period of isolation is very brief, in fact, reproductive isolation may be expected almost as a consequence of other changes. For example, Layzer (1980) has proposed a mechanism for accelerated change (cf. Section 6.2.1) which would imply this; on this model, relatively unimportant traits can change very rapidly indeed, and different changes in the two environments would have a high probability of influencing reproductive performance if the geographical isolation broke down.

In one environment, the situation is different. Maynard Smith (1966) has argued that "the crucial step in sympatric speciation is the establishment of a stable polymorphism in a heterogeneous environment", essentially through adaptation to that environment. However, the only mechanisms which appear likely to yield such a polymorphism appear to be habitat selection (Maynard Smith, 1966), and ones closely related to classical polymorphism, i.e. for a single diallelic locus, the fitness of heterozygotes must be greater than the arithmetic mean of homozygotes' fitness over all niches in the environment (Gillespie, 1976, 1977; Maynard Smith, 1980). Other models make polymorphism a consequence of very restricted conditions. For example, Stewart and Levin (1973) showed that two phenotypes could coexist in one niche on one resource, provided that their rates of increase in number depend differently on resource concentration. However, Maynard Smith (1980) showed that over a very wide range of possible resource concentrations, and a similarly wide range of selective intensities, less than 1% of all possibilities yielded a polymorphism. Stable monomorphism is also very unlikely (Griffiths, 1981). Given the infrequency of speciation in any given population's history, the rarity of appropriate conditions for sympatric speciation may be less relevant, but still the more plausible mechanism may be sought for first.

One remarkable fact to be taken into consideration is the difficulty of obtaining reproductive isolation by artificial selection within one population (Wallace, 1950; Petit et al., 1980), though much of the difficulty may have arisen, in experiments with *Drosophila*, from an incomplete understanding of reproductive behaviour (Spiess, 1970). This is in contrast to the success which has attended most attempts to increase or decrease the magnitude or incidence of a vast range of traits, many of them highly deleterious (see e.g. Lewontin, 1974, for discussion). This may have been because genes contributing to reproductive isolation are very rare in phenotypically homogeneous populations of *D. melanogaster* (Petit *et al.*, 1980) or because the genetical determination of mating behaviour is not susceptible to simple truncation selection (Sved, 1981*a*, *b*). Whatever the reason, the result is most curious.

9.2.1 prezygotic reproductive isolation

These are mechanisms which prevent the formation of hybrid zygotes, i.e. they prevent mating, fertilization or pollination. Because they are the most cost-effective form of reproductive isolation, in terms of parental investment, one might expect them to be the most advantageous in evolutionary terms and therefore to become complete rapidly. Thus, in the short-term only their breakdown may be observable.

A number of very simple models have been developed (Dickinson and Antonovics, 1973; Felsenstein, 1981 (cf. also Hutchinson, 1959); Sawyer and Hartl, 1981). These show that reproductive isolation determined by one, or a few, Mendelian genes can evolve readily if disruptive selection or precise local adaptation in at least two places occur. However, migration and recombination are very effective in preventing the development of isolation unless selection is very strong.

9.2.1.1 ecological isolation
Ecological isolation occurs where the relevant populations occupy different habitats in the same general region; it is very similar to geographical isolation. Indeed, Maynard Smith (1966) has suggested that speciation following ecological isolation through habitat selection is allopatric rather than sympatric. It could arise through a stable genetical polymorphism, if genetical variability in habitat choice existed. There is some evidence for this (e.g. Grosberg, 1981), but in general ecological isolation will be difficult to distinguish from geographical isolation unless the distribution of the species of interest is extremely uniform, which is most unlikely. Both geographical and ecological isolation allow differentiation through drift in traits of negligible selective value.

9.2.1.2 temporal isolation
By temporal isolation one means that mating or flowering times occur at different seasons, or generations are discrete and occur at different intervals.

Temporal dioecy, which occurs, for example, in a number of the Umbelliferae (Cruden and Hermann-Parker, 1977), would provide a basis for reproductive isolation of different parts of a population, e.g. along an altitudinal cline where temperature and humidity differences influence the optimal time when the anthers dehisce and the stigmas become receptive. Freeman *et al.* (1976) have presented some evidence that such factors may be important, but there is no reason to suppose that genetic drift, given geographical isolation, might not mimic the action of selection.

9.2.1.3 behavioural isolation

The term behavioural isolation applies mainly to animals, and it occurs when attraction between the sexes of different species is insufficient to cause mating. Following the idea of Wallace (1889), that reproductive isolation would evolve in the hybrid zone between distinct populations of one species, many attempts have been made to discover traits which would affect behavioural isolation in a hybrid zone. Brown and Wilson (1956) introduced the idea of character displacement, whereby in closely related populations the allopatric members would remain similar for one or more quantitative traits while the sympatric ones would diverge. This idea has proven controversial (Grant, 1975; Hendrickson, 1981; Strong and Simberloff, 1981), and what cases have been observed have largely involved traits concerned directly in reproductive behaviour (Wasserman and Koepfer, 1977, 1980; Watanabe and Kawanishi, 1979; White, 1978).

A possible example of trait modification leading to combined ethological and mechanical isolation is the geographical change in the cilia of the foreleg tibia and tarsus of male *Drosophila silvestris* (Carson and Bryant, 1979). The tibiae of the forelegs are used to stimulate the female in courtship, and are very variable in the recently evolved Hawaiian *Drosophila*. The pattern of geographical variation in these cilia is associated with inversion polymorphisms, suggesting that a trait likely to have been influenced by sexual selection is indeed associated with partial reproductive isolation in the absence of geographical isolation (cf. also Ahearn, 1980). Depending on the pattern of courtship and mating, the behavioural isolation may also approach mechanical isolation.

This example could also be explained by Muller's (1942) model, in which the mechanism of isolation has arisen as a by-product of natural selection of unknown nature in allopatry, though direct sexual selection appears less implausible. Kaneshiro (1976, 1980), however, has argued that the asymmetry of behavioural isolation which he found in four species of Hawaiian *Drosophila* is more likely to have arisen through chance or pleiotropic selection effects than direct selection, since otherwise symmetry would be expected (cf. also Wasserman and Koepfer, 1980). However, in some experimental populations of *D. melanogaster*, Markow (1981) found no relationship between mating preference and the direction of change in a number of behavioural traits when tests were conducted using males and females from selected and unselected strains. Thus, the pattern predicted by Kaneshiro was not observed (nor indeed was the very different pattern predicted by Watanabe and Kawanishi, 1979). Furthermore, Kaneshiro's argument does not appear to provide a basis for explaining unilateral incompatibility in crosses between *Lycopersicon* species where the genetics is better understood (Hogenboom, 1972*a, b*). This is because the incompatibility seems to result from the action of several different specific genes, not from those of the self-incompatibility system.

Selection, whether individual or sexual, appears to be relevant in most of the cases of behavioural isolation studied, but Nei (1975, 1976) has shown that such isolation could also occur by drift.

9.2.1.4 mechanical isolation
Physical incompatibility of the genitalia prevents copulation, that of the flower parts prevents pollination. It is difficult to imagine male and female genitalia jointly changing by mutation in such a manner as to yield distinct, compatible, novel forms incompatible with the norm. Thus, selection here appears more plausible, given as always some initial, inherited variability. A single-gene determined difference in male genitalia of East and West African strains of *Papilio dardanus* was reported by Turner *et al.* (1961). Selection in the hybrid zone between the strains, or selection or drift within the zones, could have led to the divergence, but evidence on this point is lacking.

Sexual selection might also be involved in the development of mechanical isolation, especially where the external genitalia are involved in display (e.g. West *et al.*, 1981).

9.2.1.5 pollination isolation
Specialized pollination by particular insects may preclude them from transferring pollen to different species. This mechanism is widespread, though not very frequent; co-evolution of this kind must increase the probability of chance extinction, or of the evolution of other reproductive mechanisms, because the precision of the adaptation and its joint dependence on more than one species greatly increase the potential selective advantage of a change like a structural modification where stigma and anther are physically isolated or self-compatibility in an obligate out-crossing plant (cf. Sections 5.1 and 9.2.1.6). For example, Robertson (1892) showed that the heterostylous species *Houstonia purpurea* was visited by flies, beetles and butterflies, but butterflies only pollinated the short-styled form because they could suck nectar without disturbing the anthers. Hence, flowers pollinated

only by butterflies might "result in a functional dioecism, characterised by long-styled staminate and short-styled pistillate flowers". Ornduff (1975) and Beach and Bawa (1980) have provided additional evidence for the evolution of dioecy from heterostyly in this way. Thus, an out-breeding mechanism can lead to irreversible specialization and reproductive isolation of one population of a species.

9.2.1.6. gametic isolation

External fertilization requires that female and male gametes be directed or attracted to each other; if this does not occur, reproductive isolation follows. Where fertilization is internal, the gametes or gametophytes of one species may not function in another species.

Interspecific incompatibility is the norm in flowering plants, i.e. pollen grains of one species will rarely grow far down styles of any other species; this can be determined by allelic genes (Stephens, 1946). Whatever its nature in any particular case, this mechanism interacts with the mechanism of intraspecific self-incompatibility (cf. Knox *et al.*, 1972), but as mentioned in Section 9.2.1.3, the mechanisms are likely to be genetically distinct. Furthermore, although there is evidence of selection within self-incompatibility systems (Crosby, 1940; Fisher, 1949; Williams and Gale, 1960), there is limited evidence that this is what has led to successful populations reproductively isolated by incompatibility.

What appears to be a likely consequence of self-incompatibility (i.e. intra-specific incompatibility), however, is the evolution of a large range of related mechanisms in response to the strong selection which is a result of self-incompatibility and greatly reduced numbers, i.e. when extinction is close in a self-incompatible species. Agamospermy, or asexual reproduction through seed formation, is an asexual escape from this evolutionary cul-de-sac (Marshall and Brown, 1981). However, we shall gain more insight into the evolutionary consequences by considering in some detail a number of possible modifications of self-incompatibility systems.

The simplest of these is the single-locus gametophytic system, or *Nicotiana* system, after the genus in which it was first described, where there are at least three alleles, S_i, S_j, S_k, such that a plant of genotype S_iS_j can only be fertilized by pollen of genotype S_k, $k \neq i$, $k \neq j$. The locus can be duplicated and the mechanism remains the same (see Mayo, 1978). Many other kinds of self-incompatibility exist, or can be conceived as models. Some of these could arise by mutation in the gametophytic system, providing additional flexibility and lowering the chance of extinction. First, consider Bateman's (1952) case J of incompatibility determined sporophytically by a single locus ($2Q$ is the frequency of Aa, $1 - 2Q$ is the frequency of aa):

Parents		Frequency	Offspring	
♀	♂		Aa	aa
Aa	Aa	$2Q^2$	0	0
aa	Aa	$2Q(1-2Q)$	$Q(1-2Q)$	$Q(1-2Q)$
Aa	aa	$2Q(1-2Q)$	$Q(1-2Q)$	$Q(1-2Q)$
aa	aa	$(1-2Q)^2$	0	0
			$2Q(1-2Q)$	$2Q(1-2Q)$

Here, pollen is typed sporophytically. If it were not, but the maternal phenotype remained the same (i.e. A dominant) the following situation could arise (this is a variant of Bateman's (1952) case C, where he assumed that every Aa would be self-fertilized, as it is not likely that a plant species would be expected to have evolved two out-crossing mechanisms simultaneously):

Parents		Frequency	Offspring	
			Aa	aa
♀	♂ (pollen)			
Aa	A	$2Q^2$	0	0
aa	A	$Q(1-2Q)$	$Q(1-2Q)$	0
Aa	a	$2Q(1-Q)$	$Q(1-Q)$	$Q(1-Q)$
aa	a	$(1-Q)(1-2Q)$	0	0
			$Q(2-3Q)$	$Q(1-Q)$

The frequency of Aa in the offspring is given by

$$2Q' = \frac{Q(2-3Q)}{Q(2-3Q)+Q(1-Q)}$$

$$= \frac{4-6Q}{2(3-4Q)}.$$

When the population reaches equilibrium, i.e. $2Q' = 2Q$,

$$2Q \simeq 0{\cdot}61$$

(cf. Bateman's (1952) value of 2/3, obtained by assuming that every Aa is self-fertilized).

Instead of sporophytic determination arising by a mutation at the self-incompatibility locus, such a system might evolve from a gametophytic system by the acquisition of dominance, whether for the determination of male specificity, female specificity, or both, by mutation at another locus.

Consider a perennial species, with the *Nicotiana* incompatibility system. Now suppose that by chance it has been reduced to two alleles, A_1 and A_2.

Then all the plants are A_1A_2, and the species will die out if no new evolutionary step is taken.

Suppose there is a locus M affecting female specificity so that A_1A_2M- acts as if it were A_1A_1. Then further suppose that seeds bearing M have greatly lowered viability, both in germination and early growth, but that a plant bearing M which reaches maturity suffers no impairment compared with mm plants. Let the mutation rate from m to M be μ, and the relative viabilities have the proportions:

$$
\begin{array}{ccc}
MM & Mm & mm \\
1-s & 1-s & 1
\end{array}
$$

Then the frequency of M, disregarding its influence on the other locus, is about μ/s. In other words, if s is substantial, M will be very rare, and will not be likely to alter the mating system in any substantial way, if there is competition between young plants. But if the only seed being set is by plants carrying M, as will be the case in the situation postulated, the higher mortality among M plants will not matter. Further, as each new generation grows from the selfed A_1A_2M-, the frequency of M will rise. The following genotypes will be present:

	A_1A_2			A_2A_2		
	MM	Mm	mm	MM	Mm	mm
♀ phenotype	A_1	A_2	A_1A_2	A_2	A_2	A_2

Thus the situation is the same as if there were initially three female genotypes, A_1A_1, A_1A_2 and A_2A_2. Whether M is fixed or not, this is an ineffective out-breeding system, as noted earlier, since some plants will be self-fertile.

If, on the other hand, M acts to make $A_1A_2 A_1$-specific in style and pollen, the system which develops is Bateman's type J, which is stable, thereby assuring out-crossing. As Bateman (1952) indicated, it might well have been ancestral to heterostyly, but this is a very difficult hypothesis to test.

It can readily be seen that if a mutation occurs which will, say, make A_1 dominant to other alleles in both male and female specificity, it will not necessarily be at an advantage initially because a plant of phenotype A_1, while it can be pollinated by all except A_1 pollen, will only be able to pollinate non-A_1 individuals. Thus the new mutant may not be able to spread, as the plants near the one carrying the new mutant will be more likely than those in the population at large to bear A_1 (see Charlesworth and Charlesworth, 1979).

Since even the loci involved in gametophytic incompatibility are complex (Lewis, 1960, 1965; Pandey, 1970), evolution of additional functions in the manner suggested here is not implausible. A gene with an allele having many of the required properties was found in *Petunia hybrida* by Mosig (1960). Charlesworth and Charlesworth (1979) and Charlesworth (1979) have studied such problems in considerable detail; my intention here has been simply to indicate the kinds of events which could lead to divergent reproductive systems rather than extinction. A more general conclusion may be drawn, or at least supported: without knowing what, if any, mutants are present in a population confronted with a novel selection pressure, one cannot predict what direction evolution will take. That an explanation may be provided for events which have occurred but not predicted for those yet to occur is, of course, one of the well-known differences between the biological and phsyical sciences.

9.2.2 zygotic isolation

These mechanisms reduce the viability or fertility of hybrid zygotes. As implied in Section 9.2.1, one might expect, *a priori* that zygotic mechanisms, which inherently carry an initially higher reproductive cost, might take longer to establish and therefore be observable in more stages at any particular time. Darwin (1859) recognized their importance and devoted considerable attention to the sterility of hybrids and the taxonomic problem of apparently distinct species which could be crossed successfully. What I am concerned to do here, however, is simply to illustrate the types of isolation and their implications for the action of selection rather than for taxonomy.

Early work on mechanisms of zygotic isolation was mainly concerned with the fixation of sets of genes having properties similar to self-incompatibility (cf. e.g. Dobzhansky, 1937; Muller, 1942). However, more recently it has been recognized that because chromosomal mutations such as balanced translocations make a very effective post-mating barrier to population mixing, they may well have been important in evolutionary terms (White, 1978). If this was the case, then unless assortative mating occurred, the population in which balanced translocations were fixed must have been very small (Wright, 1941). In this case, random factors would have been more critical than selective factors.

9.2.2.1 hybrid inviability
Hybrid zygotes are either inviable or have reduced viability.

A gene, *Lhr*, found in a natural population of *Drosophila simulans* (Watanabe, 1979) provides a good example of this process. This gene altered the inviability of hybrid progeny between *D. simulans* and *D. melanogaster*. The progeny produced in different crosses were as follows:

D. melanogaster female × *D. simulans* male   *D. simulans* female × *D. simulans* male
Lhr Absent   female only   male only
Lhr Present   50% female 50% male   14% female 86% male

All hybrids were sterile, whether *Lhr* was present or absent. Thus, the extent of reproductive isolation of the two species depends on both hybrid sterility and hybrid inviability. It will indeed be rare that the two phenomena are not associated, since the latter implies the former.

9.2.2.2 hybrid sterility

The F_1 hybrids of at least one sex fail to produce functional gametes. Oka (1957, 1974) showed that hybrid sterility in rice could arise through the interaction of complementary dominant lethal factors, and similar mechanisms have been found in other plants. In addition, combining the cytoplasm of one plant species with the nucleus of another may result in male sterility (see Edwardson, 1970, for a review). Thus, reproductive isolation may involve what is essentially variation between clones as well as variation through Mendelian genetic changes. The dynamics of such clonal variation is not well understood, exept at the simplest level, though recent developments in the genetics of organelles (cf. Takahata and Maruyama, 1981) may be applicable. If so, drift is likely to have been important in the origin of hybrid sterility.

As noted in Section 9.2.2.1, hybrid sterility is a less extreme form of reproductive isolation, and in plants may still yield significant populations of hybrids if these can reproduce vegetatively. Hence, it may be expected that selection will tend to move the hybrids from sterility to inviability unless they have some peculiar advantage.

It is therefore remarkable that stable, narrow hybrid zones between related taxa are not infrequent (Anderson, 1948; Dobzhansky, 1970). Various models have been proposed to explain their persistence, including hybrid superiority in the narrow zone with inferiority elsewhere and differential selection in the two basic habitats. In particular cases, some hypotheses have been rejected. For example, Bull (1979) showed that hybrid inferiority was not the case for the hybrid zone between two Australian frogs, *Ranidella insignifera* and *R. pseudoinsignifera*. Dobzhansky's (1970) general explanation was that these zones are transient and will disappear, so that directional selection is the key.

9.2.2.3 hybrid breakdown

The F_2 or back-cross hybrids have reduced viability or fertility. Hybrid dysgenesis in *Drosophila melanogaster* (Hiraizumi, 1971; Picard and L'Héritier, 1971; Kidwell *et al.*, 1977; Sved, 1979) is a widespread phenomenon which incorporates elements of hybrid sterility and hybrid breakdown. The initial observation was that in certain crosses, e.g. between some laboratory strains and some wild-type strains, male recombination and an increased frequency of mutation occurred. It was found to be non-reciprocal, involving a cytoplasmic contribution from the mother as well as chromosomal contributions from both parents.

The mechanism is not completely clear (Sved, 1978, 1979), so that speculation on its evolutionary relevance is hazardous: "it would be fortuitous if hybrid dysgenesis provided a model of the changes that normally accompany speciation" (Sved, 1979). None the less, it is a mechanism of reproductive isolation between distinct strains which is relatively widespread in one species of *Drosophila*; it would be of interest to learn of its existence in the very recently diverse *Drosophila* fauna of the Hawaiian islands, where selective factors, if relevant, may still be identifiable.

TOPICS FOR DISCUSSION

1. In a long-term study of inbreeding in a small population of the great tit, *Parus major*, van Noordwijk and Scharloo (1981) found good evidence of inbreeding depression for egg-hatching, but found, none the less, that overall reproductive success of inbred individuals was higher than non-inbred individuals, partly because of the very great success of a few (presumably superior) individuals. Is this typical of the circumstances that, in general, will make inbreeding advantageous in a naturally out-breeding species?
2. Does the "cost of meiosis" differ between outbreeding hermaphrodites and male haploid–female diploid organisms? (Charlesworth, 1980).
3. Are departures from the hermaphrodite state like dioecy, monoecy or andromonoecy more likely to lead to reproductive isolation of different populations than self-incompatibility mechanisms?
4. Is reproductive isolation a necessary precursor of speciation?

10   speciation

10.1 INTRODUCTION

Before considering the special relevance of natural selection to speciation, as opposed to changes within a lineage, we should consider briefly what a species is. Dobzhansky (1970) put it well:

> A species, like a race or a genus or a family, is a group concept and a category of classification. A species is, however, also something else: a supra-individual biological system, the perpetuation of which from generation to generation depends on the reproductive bonds between its members.

In most cases, then, a species may be regarded as more than the taxonomic convenience which Darwin sometimes suggested it was. However, controversy will still arise in defining a species, since reproductive isolation is not necessarily as cut and dried as definition may require (cf. e.g. Clarke and Sheppard, 1970).

I should also note that mechanisms for speciation based not on biological processes but on more general physical considerations, while they may be important, have as yet shed no insight on speciation. Here I shall therefore not consider such models as the catastrophist (Dodson and Hallam, 1977) nor the closely related phase change (Ruelle, 1981).

Nor shall I enter the difficult area of construction of and inference from phylogenies. Osborn (1909) wrote that Darwin *"never observed a single phylum"* (Osborn's italics). Whatever the truth of this remarkable statement, here I shall not observe any in the sense of deriving them, nor enter into any of the important phylogenetic controversies (Hennig, 1966; Mayr, 1974; Felsenstein, 1978b; Harper, 1979; Platnick, 1979; McMorris and Zaslavsky, 1981; Robinson and Foulds, 1981; Jensen and Barbour, 1981; Astolfi *et al.*, 1981). I have, however, obviously made extensive use of results based on phylogenetic analysis, especially in Chapter 6.

As Mayr (1974) noted, Darwin (1859) was the first to point to the two essentially different processes of speciation: reproductive isolation, or splitting, and subsequent ecological differentiation which would minimize competition in sympatric populations but which could also occur by chance in allopatric populations. In effect, Darwin had recognized the competitive exclusion principle, i.e. that no two species can coexist indefinitely when they are limited by the same resource (Volterra, 1928, Hardin, 1960; see Armstrong and McGehee, 1980, for a review of the applicability of this principle). Selection, in Darwin's view, was important in both processes of speciation.

10.2 TYPES OF SPECIATION

The splitting of one species into more than one, however it comes about, seems always, though uncertainly, to create controversy. Much of our understanding of the effects of selection relates to changes in traits, both under artificial and natural selection in well-defined populations, and over geological time in populations whose breeding structure is no longer determinable. Osborn (1915) suggested that

> The old and ever vague problem of the origin of species is being resolved into the newer and more definite problem of the origin of characters; in the dim future when we know how and why new characters originate, and how and why they transform and disappear, the problem of species will have long been solved and well-nigh forgotten.

This optimistic prediction has yet to be fulfilled, but there are more recent related ideas which may be tested. Wright (1967), as has frequently been emphasized by Gould (e.g. 1977), made a most important suggestion with respect to directional change and speciation, namely that speciation might well be random with respect to long-term evolutionary trends, in the same way that mutations might occur at random with respect to short-term evolutionary trends. If these ideas have any validity, and they have wide support (Stebbins and Ayala, 1981), we may expect a range of processes to contribute to speciation.

10.2.1 polyploidy

10.2.1.1 occurrence
While speciation through polyploidy must be sympatric, it is considered separately because it is such a distinct mechanism.

The process, which occurs very largely in plants, is that two species which have been geographically isolated cease to be so for some reason and pollination is effected and followed by genomic balancing to restore fertility. Two possible mechanisms are described in Section 10.2.1.3.

Hennig (1966) has criticized the description of these processes as speciation in that species hybridization means that reproductive isolation cannot have been complete. However, in the sense that it produces sterile hybrids, and that the polyploidy which restores fertility renders the offspring species reproductively isolated from both parent species, it meets most normal definitions of species. (Hennig also criticized the description of the process as speciation because in his system of taxonomy, cladistics, the polyphyletic origin of species was defined not to exist.)

While polyploidy has been a very important mechanism for speciation (Stebbins, 1971), it is largely restricted to plants, invertebrates and other non-amniote animals. The requirement of crossing for reproduction in bisexual animals is probably the reason for the limited range of polyploidy in animals (White, 1973). Chromosomal sex-determination mechanisms are so liable to disruption on polyploidy that where they exist, polyploidy is not to be expected (Muller, 1925; Ohno, 1970).

10.2.1.2 consequences

Nuclear DNA content in plants influences many components of nuclear and cellular size and mass. It also influences duration and rate of important developmental processes including mitosis and meiosis (Stebbins, 1949; van't Hof and Sparrow, 1963; van't Hof, 1965; Leps *et al.*, 1980). Minimum generation time is thereby directly associated with nuclear DNA content, and Bennett (1972) has suggested that this is not only the result of information carried in the DNA but also because of the "physicomechanical effects of its mass". If this is the case, then it is a constraint on the amount of DNA which can be accumulated.

B chromosome accumulation appears to be a case in point. It has long been recognized (e.g. Darlington, 1956) that these supernumary or accessory chromosomes, which vary in number among individuals of a species and may be entirely absent, may be disadvantageous when present in large numbers. They generally reduce fertility by disrupting mitosis in gametogenesis, but in fact may also be advantageous; in tetraploid *Lolium remotum* × *Lolium perenne* hybrids, they alter pairing from autotetraploid with multivalents to regular allopolyploid with bivalents (Rees and Hutchinson, 1973). It is difficult to predict the effects of B chromosomes. For example, under highly competitive, i.e. crowded, conditions, B chromosomes are disadvantageous in rye (*Secale cereale*) but advantageous in perennial ryegrass (*Lolium perenne*) (Rees and Hutchinson, 1973). The origins of B chromosomes are uncertain (Hewitt, 1973), but they certainly illustrate how accumulation of DNA is not without direct, though non-Mendelian effects.

As Haldane (1926) was the first to show, change in response to natural selection will be slower in polyploids than in diploids, in general (cf. also Mayo, 1971; Hill, 1971; Honne, 1979). For a given size of population, polyploids will be more resistant to gene frequency change by drift than diploids (see e.g. Mayo, 1971), and will also allow higher equilibrium frequencies of deleterious recessives than diploids (Bennett, 1976). Polyploids may therefore be regarded as being, in genetical terms, better buffered than diploids, their more sluggish response to the directional selection being a corollary of this.

The relationship between genetical variability and ploidy is not simple. Table 10.1 provides some illustrative data.

Because polyploidy offers the opportunity for divergent enzyme evolution as well as for gene silencing (Section 7.1), it is the most direct way in which enzymic diversity might be expected to occur in plants, as opposed to animals. There is no evidence on this point as yet.

Table 10.1 Genetical variability and ploidy.

Organism	Ploidy	Heterozygosity of polymorphic loci	References
Sexually reproducing			
Drosophila pseudoobscura	1½	0·22 ± 0·11	Cooper et al. (1979)
D. pseudoobscura	2	0·18 ± 0·03	Cooper et al. (1979)
D. persimilis	1	0·35 ± 0·09	Cooper et al. (1979)
D. persimilis	2	0·22 ± 0·06	Cooper et al. (1979)
D. robusta	1	0·16 ± 0·07	Cooper et al. (1979)
D. robusta	1	0·21 ± 0·04	Cooper et al. (1979)
Hyla chrysoscelis	2	0·07*	Ralin and Selander (1979)
H. versicolor	4	0·32*	Ralin and Selander (1979)
Marsupials	1½	0·04 ± 0·01	Cooper et al. (1979)
(10 spp. summarized)	2	0·04 ± 0·02	Cooper et al. (1979)
Largely asexually reproducing			
Myzus persicae	2	no loci polymorphic	Wool et al. (1978)
Solenobia triquetrella	2	0·70†	Lokki et al. (1975)
S. triquetrella	4	0·31†	Lokki et al. (1975)
Daphnia carinata	2	0·01	Lokki et ai. (1975)

* All loci, including monomorphic loci.
† Only in samples containing heterozygotes.

As noted earlier, sexual reproduction may be disturbed by polyploidy. Thus, many successful polyploids are asexual in reproduction (Stebbins, 1971). Their success in evolution may reflect their improved ability in colonization from a few founding plants. A classical example is the polyploid *Spartina townsendii*, which resulted from the cross of the European *S. maritima* and the American *S. alternifolia*, replacing *S. maritima* along the coast of France (Huskins, 1930; Zeven, 1979).

10.2.1.3 mechanisms
Autopolyploidy may arise as a failure of the reduction division of meiosis, or as a result of suppression of mitosis after chromosome replication. Allopolyploidy, however, occurs through a more complex process, in that it

requires a cross between organisms having different chromosome complements. The two proposed mechanisms are as follows.

In the first, the chromosomal complement of the sterile hybrid $(2x)$ resulting from the union of ordinary reduced gametes (x) is doubled, and the resultant polyploid $(4x)$ is fertile and reproductively isolated from both parental species. However, as de Wet (1979) has emphasized, spontaneous doubling is rare, and therefore this process is likely to be very rare. Polyploidy via cytologically non-reduced gametes is therefore more common. It may be a two-step process. First, a diploid female gamete $(2n)$ is fertilized by a haploid male gamete (n) to produce a triploid $(3x)$ which gives nonreduced female gametes $(3n)$. These in turn are fertilized by haploid male gametes (n) to give a tetraploid $(4x)$.

With the recognition that single genes (Riley, 1960) as well as B chromosomes, mentioned above, may change pairing from polyploid to diploid, it is evident that the evolution of the meiotic and mitotic mechanisms is relevant to the role of ploidy in evolution. However, the direct role of natural selection is not clear.

10.2.1.4 major evolutionary departures
Apart from individual cases of speciation through polyploidy, many of which are well-documented (Stebbins, 1971), there is also the possibility that major evolutionary changes have required very substantial changes in DNA content. Ohno (1970) has strongly emphasized this point in his theory of evolution by gene duplication.

Sparrow and Nauman (1976) have taken this idea further, and have suggested that a plot of the distributions of the logarithm of nucleic acid contents per genome against the powers of two displays a highly informative periodicity, so that divergence of major phylogenetic groups of organisms is related to doublings of nucleic acid content, possibly independent of ploidy. This latter conclusion derives from the linear relationship between the minimum nucleic acid content for each major phylogenetic group and the "genome doubling sequence" (powers of two). The inference from these results is that a major evolutionary departure requires a major change in the quantity of genomic nucleic acid. Why this should be so is not clear, nor is its implication for the role of natural selection at all obvious. Furthermore, it does not aid in the solution of the problem of widely differing DNA content in closely related species nor why, for example, gymnosperms should have approximately four times as much DNA per genome as angiosperms.

10.2.2 sympatry

Speciation within one environment was regarded by Mayr (1963), on the basis of a great deal of evidence, as being so rare as to be negligible. This is in the case of animal evolution. However, models of sympatric speciation have been suggested by, among others, Fisher (1930), Maynard Smith (1966) and James (1970), and apparent sympatric speciation has also been observed in progress (Murray and Clarke, 1980). Murray and Clarke (1980) examined nine taxa of land snails belonging to the genus *Partula*, and found as many as four in a single locality with no evidence of hybridization. They also found two cases of circular overlap, i.e. so-called ring species, and concluded that all the taxa which they examined were likely to have been the descendants of a single colonization, with successive adaptation to a variety of similar but distinct unoccupied niches. Strictly, sympatry requires complete overlap, with the same niches occupied by the two populations of interest. Nevertheless these results display a process very close to the definition of sympatric speciation, since niche identity is also very hard to demonstrate.

It is only in the case of very rapid recent speciation that this kind of investigation is possible. Therefore, most of the evidence relating to it reflects patterns of colonization. This need not mean that it is a rare event in evolutionary terms.

10.2.3 allopatry

Allopatry may be regarded as the norm for animals. The most widely accepted model is that of Mayr (1963), whereby small, isolated populations become distinct both by local adaptation, i.e. natural selection, and by genetical drift, this quantitative change gradually becoming large enough to lead to a qualitative change, i.e. a "genetic revolution", yielding a new, distinct species. Lande (1980b) has provided evidence that the "founder effect", i.e. the influence of chance in the initial genetical composition of the small, isolated populations, is less important than suggested by Mayr, essentially because of the great variability of quantitative traits and the strength of natural selection, as measured by the methods discussed in Section 6.4.

Templeton (1980b) has pointed out that Mayr's theory accounts for the possibility of speciation, but "fails to predict which founder events lead to speciation and which lead to trivial changes from a cladistic point of view". Templeton has suggested that certain properties of founder populations make speciation more or less likely: those shown in Table 10.2.

About half the characters relate to reproductive performance and population structure, a few to reproductive behaviour and a few to the genetic architecture of the species.

The attributes relating to reproductive behaviour appear likely to be unstable, through conflicts between carrying capacity and population growth (Felsenstein, 1979). Thus, under the conditions which favour an increase in speciation rate, an increase in extinction is also to be expected.

Table 10.2 Characteristics of founder populations that influence the chance of speciation.

Attributes that increase the chance	Attributes that decrease the chance
Average number of offspring large	Average number of offspring small
Reproductive value of founders high	Reproductive value of founders low
Open niche allowing population flush	Population flush not possible
Initial density low	Initial density high
Initial subdivided population structure	Initial panmictic population structure
Overlapping generations	Discrete generations
Assortative mating	Disassortative mating
Sexual selection on the mate recognition system	Rare male or similar sexual selection
Imprinting, partially learned sexual behaviour	Sexual behaviour totally genetic
Chromosome number large	Chromosome number small
Total genomic map length large	Total genomic map length small
Cross-over suppressors few and easily lost	Cross-over suppressors many and not easily lost

From Templeton (1980*b*).

10.2.4 peripheral isolates

Following the suggestions of Wright (1931, 1949), that evolution proceeds most rapidly in small, partly isolated populations, it has been suggested (see e.g. Cracraft, 1974; Carson, 1975; Gould and Eldredge, 1977), that this is where most speciation will occur. One implication is that transitional fossil forms will be very rare, both because speciation is rapid and because it occurs in small populations relative to the ancestral species.

It should be noted, however, that the subdivided population need not undergo change solely in peripheral isolates, but in any of the partly isolated sub-populations which constitute the whole (cf. Templeton, 1980). Furthermore, it is very hard to predict what will happen in peripheral populations. It is of great interest that the butterflies colonizing Australia, 30 million years ago,

> which might be considered outliers of genera well developed elsewhere, remain feeding on plants no different from those found in the main centres of distribution of the genus and they have not moved on to distinctively Australian plants.
>
> (Symon, 1980)

This is a clear case, on a very large scale, where peripheral isolation and colonization have not led to rapid divergence and speciation.

10.3 RATES OF SPECIATION

Although Darwin (1859) did not say that speciation must proceed by anagenesis, i.e. change within a lineage, rather than by cladogenesis, or splitting, and though he did not say that speciation must inevitably be an almost imperceptibly slow process, it has been claimed that slow anagenesis has been widely accepted as the Darwinian norm (cf. e.g. Eldredge and Gould, 1972). It is therefore worth noting that this view has by no means been the norm for a very long time, if ever. For example, Anderson and Stebbins (1954) wrote

> It has been established by recent work in Palaeontology and Systematics that evolution has not proceeded at a slow even rate. There have instead been bursts of evolutionary activity as for example when large freshwater lakes (Baikal, Tanganyika, and Lanao) were created *de novo*.

Anderson and Stebbins concluded that this rapid evolution, by which they largely meant speciation, had as its stimulus hybridization when "diverse faunas and floras were brought together" in different "habitats where some hybrid derivatives would have been at a selective advantage". Others (e.g. Simpson, 1953; Mayr, 1963) have emphasized colonization of new environments or changes to marginal populations as the major relevant stimuli. Evidently, major environmental changes such as the end of an ice age would also constitute the kind of environmental opportunity which *a priori* should lead to increased rates of speciation. Dispersal ability will of itself influence the rate of speciation. If an organism has low dispersal ability and inhabits an environment which shows little change, it has little opportunity for divergence, if divergence has any adaptive significance, and may be expected to have a low rate of speciation. This is true, for example, of soil nematodes.

Much work has been carried out since Simpson (1949) and Haldane (1949) suggested how evolutionary rates might be investigated systematically. Simpson then gave as examples 7500 years for a subspecies of a particular deer to arise in Scotland and 0·15 genera per million years in the case of the horse. Speciation may be much more rapid as in the Hawaiian *Drosophila*, or fantastically slow, as in the horseshoe crab, *Limulus*, which appears to have persisted, physically unchanged, for 200 million years (Mayr, 1963).

Bookstein *et al.* (1978) have examined two fossil lineages, a primate, *Pelycodus* (see Section 6.4), and a condylarth *Hyopsodus*, for evidence of distinguishable rates and rapid rate changes. On their interpretation, of 17 definable cases, 12 showed gradual change, four sharp changes, and one no change. This interpretation, however, depends to some extent on arbitrary divisions between strata. Godfrey and Jacobs (1981) have emphasized just how difficult it is to discriminate between models using fossil data, and given the very variable variability of the material, almost any model could fit. Large numbers of similar analyses will be needed before conclusions can be expressed with any confidence.

Explanations for major changes in speciation rates have frequently been sought in terms of major climatic changes, as in the onset of the age of the mammals, when such rates varied upwards from about 3 to about 100 events per potential ancestor per million years (van Valen, 1978). It has further been suggested that the major extinctions over roughly the same period were, at least to some extent, occasioned by the explosive radiation of the mammals (van Valen and Sloan, 1977). These extinctions will be considered further in Section 10.4.

As we have seen in Chapters 2 and 6, both time and selective processes are sufficient to account for all rates estimated as yet. Even if speciation is usually very rapid in geological terms (though not as rapid as speciation through polyploidy), it is still a slow process in genetical terms. However, explanations of major changes in pattern are in general no more than historical in nature.

10.4 EXTINCTION

It has been said that extinction is the hardest fact of evolution to account for in terms of natural selection (cf. e.g. Løvtrup, 1975). In a sense, this has some truth, but overall it is quite wrong; Darwin would certainly have disagreed (cf. Mayo and Bishop, 1979). The general explanation is that the genetical variation available in a species was not sufficient to meet an environmental challenge or it was unable to disperse, to escape that challenge. I have already alluded in another context to the death of the dodo; it is one of a number of flightless birds extinguished by colonization of a new environment by mammalian carnivores. Re-evolution of flight (contrary to Dollo's "law") or a development of great body size and/or extreme leg length would appear to have been the paths of escape, but time was insufficient. The dodo had occupied a niche made available by the absence of medium-sized and large mammals, but since it had evolved also in the absence of predation, it was optimally adapted only for its pre-colonized environment – if indeed it was in any sense optimal. But the sense in which the criticism has merit is that it is not necessarily possible to predict extinction. If the environment is always deteriorating (cf. Chapter 6), then the general prediction is that all species will be extinguished. This is not useful in any particular case. In the case of the dodo, and many flightless birds in New Zealand, one could make the prediction that they will all be extinguished in the event of no specific action to preserve them artificially (cf. e.g. Diamond and Veitch, 1981). The same argument applies to large carnivores and venomous snakes in highly populated areas, just as it does to highly specialized relic species in single environments.

In Section 9.2.1.6, it was shown that in a self-incompatible population a number of different mutational events could lead to the avoidance of speciation, through apomixis, self-fertility or a change in the type of self-incompatibility. This highlights the unpredictability of the events associated with the extinction of a single species.

Major colonization will always result in some extinction; only the most obvious cases will be predictable. A case of great interest, cited earlier, is the failure of Australian butterflies to take advantage for larval feeding of the most widely distributed plant families in Australia, such as the Casuarinaceae, Proteaceae, Chenopodiaceae, Myrtaceae and Goodeniaceae (Symon, 1980). Thus, in the 25–35 million years available, co-evolution has been very slow. Evidently the genetic variability within the butterflies has been inadequate or the larva–plant associations already established when colonization by butterflies began were extremely strong. A general inadequacy of adaptation seems to be what has accompanied colonization. Implications for extinction are most uncertain, except for the trivial one that extinction would be more probable for any particular taxon than if that taxon had diversified its host plants widely.

To clarify further the problem of the unpredictability of extinction, consider a major series of extinctions, those at the Cretaceous—Tertiary boundary.

The Cretaceous—Tertiary boundary is a very precise change in the stratigraphic record which occurred about 65 million years ago (see Pearson, 1978, and Smit and Hertogen, 1980, for references). It has been suggested that an enormous range of species of Foraminifera and other aquatic organisms were extinguished within 200 years (Herm, 1965). This was followed by a remarkable diversification from the remnants of these groups.

So much is hardly in dispute, though Archibald (1981; see also Clemens and Archibald, 1980) has pointed out a number of departures from rapid extinction, at least as regards terrestrial fauna. (Most major extinctions have affected larger, more specialized species more than other species, as emphasized by Boucot (1976), but this pattern also is not universal.) The cause of the major extinctions is certainly not clear. At least six "gradual" explanations have been suggested, as well as four "catastrophic" possibilities. All appear to ignore the fact that not all relevant groups disappearing at about this time follow the same abrupt pattern, nor do they even all begin to disappear at the same time (Wiedman, 1969). The first key element to note, however, is that the suddenness of the events is such that natural selection had little time to act. The second point is that even if the cause was, for example, a cyclical variation in some climatic factor (see Pearson, 1978), without knowing what that factor is and how it cycles, if it still does, one can make no predictions based on its state at any particular time. This does not mean that climate-related predictions cannot be made: they can but they can be tested historically only, given our impatience and eagerness to know all before death.

In order to consider how convincing the tests may be, we consider in a little more detail two of the possible causes of the Cretaceous extinction.

Alvarez *et al.* (1980) have suggested that the cause was an extra-terrestrial event, probably the impact of an enormous asteroid, since there are anomalous levels of various metallic elements in the relevant strata. These results have been confirmed and extended by others, such as Ganapathy (1980) who has shown that noble metals are enriched in a Cretaceous–Tertiary boundary clay, and who had already demonstrated that such enrichment is a "sensitive indicator of meteorites".

The mechanism of extinction is postulated to have been the immense cloud of debris thrown up by the impact of an asteroid about 10 km in diameter (Alvarez *et al.*, 1980). This would have blocked out the sun for sufficiently long to prevent photosynthesis "and thus attack food chains at their origins". Coccolith-producing algae, at the end of the food chain, might therefore be expected to be extinguished very rapidly, which does appear to have been the case. Similarly, very large, terrestrial animals, at the top of food chains, might also have been extinguished rapidly. Indeed they were extinguished, but probably not nearly as rapidly (van Valen and Sloan, 1977). Land plant changes were also gradual (Hickey, 1981).

Gartner and co-workers (Gartner and Keany, 1978; Gartner and McGuirk, 1979) have concluded that the rapid extinction of the marine biota and slower extinction of the large terrestrial mammals are more precisely consonant with the hypothesis that the Arctic Ocean became isolated largely from the other major oceans in the late Cretaceous, through "tectonism and a gradual regression of shelf seas". Then the Arctic Ocean was diluted by rain and inflow, so that when Greenland and Norway parted 65 million years ago, lighter, fresher water flowed into the major oceans, covering the surface with a layer of lower salinity, thereby extinguishing surface dwellers and also cutting off oxygen to deeper dwellers. The oceanic cooling would have had a slower effect on terrestrial life than on marine life, thereby accounting for the slower extinction of the large animals.

Further investigation will cause these and other relevant theories to be further modified or abandoned; here I merely conclude that models of speciation, gradual or irregular, are not appropriate to explain major extinctions.

10.5 PATTERNS OF SPECIATION AND EXTINCTION

The recognition of the fact of widespread extinction and the concept of evolution by natural selection jointly led to the idea that species or even higher taxa might exhibit life cycles analogous to those of individuals. The idea became particularly popular through considerations of the fate of man (e.g. Spengler, 1919). However, there was also more objective support from the fossil record. It has frequently been found that the appearance of a new type is followed by "explosive radiation" into a substantial range of forms based on the novel type. This increase in the number of forms eventually ceases, and then in most cases is followed by a slow decline until the entire clade disappears.

Inherent in the interpretations of such patterns presented by, for example, Schindewolf (1945, 1950), is the idea that in the last phase, which Schindewolf called typolysis, the remaining species are degenerate and at an evolutionary dead-end. The classical example is that of the heteromorph ammonoids which were mainly extinguished at the Cretaceous–Tertiary boundary. However, Wiedmann (1969) has shown that the forms still prevalent at the end of the Cretaceous were by no means degenerate types, and in fact in some lineages, Dollo's rule appeared to have been disproved, i.e. tightly coiled ancestral forms appear to have evolved from uncoiled forms. We have already seen how the Cretaceous–Tertiary extinction has evoked many, largely exogenous explanations, and that as yet none is completely satisfactory. It does not seem, however, that the endogenous explanation of Schindewolf and others is satisfactory either.

In an attempt to detect whether the observed patterns are real, or artefacts arising from random processes, Gould et al. (1977) have constructed a Monte Carlo model of the production of evolutionary trees by lineage branching, branching and extinction being equally likely. They then divided these trees into clades and compared their forms with those patterns apparent in the fossil record. The real and simulated patterns were similar, though with some important differences, such as surviving clades apparently being larger than extinct clades. There may be elements of artefact in both the similarities and the differences, depending how the simulation models were built and how well the fossil record reflects extinct clades as compared with surviving clades, but none the less these results support the view that typolysis is an unnecessary idea.

The results of Gould and co-workers appear to imply that speciation following immigration rapidly reaches some maximum; the existence of such maxima, results of unknown environmental constraints, may be one type of pattern which warrants further exploration, given this empirical support. In fact, two broad patterns of rapid origination and extinction have been predicted; one which suggests that major adaptive radiation begins with early diversification of major taxa as original ancestors occupying major, new, adaptive zones, followed by later diversification at a minor level "filling the gaps" (Valentine, 1969), the other suggesting that diversity in an adaptive zone rises to a saturation point without necessarily having any change in the pattern of diversification (MacArthur and Wilson, 1963; Mark and

Flessa, 1977). According to Cifelli (1981), the former implies a negative association between rates of origination and extinction, the latter a negative association. Cifelli presented evidence on change in two ungulate orders from the Eocene to the Pleistocene which appeared to show limited support for the former model described above, a weak positive association being apparent between numbers of originations and extinctions. Apart from any problems generated by the nature of the fossil record (Raup, 1972), however, there are grounds for arguing that origination and extinction may be associated in the absence of any causal mechanisms. Although the relationship may not be linear or simple, it is evident that while large or small numbers of extinctions can occur among a large number of taxa, among a small number of taxa only a small number of extinctions can occur, and if one considers rates rather than numbers, a high rate of origination from a vey diverse fauna requires an enormous absolute number of originations.

This does not mean that the idea of a taxon cycle is totally useless; as a description of the pattern following migration of a small number of organisms to a new environment, it is simple and clear. For example, Ricklefs and Cox (1972) have considered such cycles in birds in the West Indies. The pattern of immigration from the mainland to the islands, followed by adaptive radiation and later extinction, describes satisfactorily what seems to have happened. It does not, however, provide any additional insight into evolution above the species level.

TOPICS FOR DISCUSSION

1. What evidence would demonstrate the critical importance of polyploidy in major evolutionary departures?
2. What are the most probable evolutionary effects of polyploidy?
3. Is polyploidy likely to have been a less frequent mechanism of evolutionary change in Recent than in Pre-Cambrian times?
4. In what ways are the evolutionary consequences of polyploidy distinct from those of duplication?
5. Is character displacement irrelevant in speciation?
6. How would one demonstrate that speciation was random with respect to the direction of evolutionary change in some quantitative trait?
7. In what ways has population genetics contributed to our understanding of speciation?
8. If, for any major taxonomic group, taxa have a relatively constant rate of extinction over a long period, does this result provide evidence for selection as the major determinant of extinction?
9. Are organisms like *Limulus*, *Sphenodon* and *Didelphis*, all of which appear to have changed little in morphology in 70–200 million years, "animals for which evolution essentially stopped long ago"? (Simpson, 1953b).

11 conclusions in terms of historical science

In this book, I have attempted to assess whether the processes of natural selection, while not uniquely responsible for "descent with modification", as Darwin (1859) put it, are sufficiently powerful to account for adaptation within the constraints provided, on the one hand, by the existence of a real world and, on the other hand, by the penalties imposed by success. Changes within one population can be well-described by the effects of natural selection and chance on the variability generated by mutation, maintained by population numbers, and exposed by segregation and recombination. Both change and adaptation, however, as I have shown, are not readily predictable. I have also described how well the phenomena of extinction, some of the most certain phenomena in biology as understood today if only through the intervention of man, are explained in terms of natural selection.

I have reviewed some of the evidence, anecdotal, historical, observational and experimental, for the existence of different types of natural selection, stabilizing and directional. In doing so, I hope to have shown that the "story-telling" mode of evolutionary explanation is useful. Figure 11.1 shows some of the changes postulated to have occurred in the evolution of man. Figure 11.1 is a diagram of a set of plausible hypotheses, which can readily be extended. For example, it does not include the extension of the period of parental care which may well relate to increased brain size, nor does it contain the evolution or development of speech, the biological content of which remains highly controversial (Chomsky, 1976; Hansen, 1981). Furthermore, given this framework, contrary hypotheses may be set up. For example, take Cachel's (1979) account of the evolution of manual dexterity in higher primates:

> Perhaps the development of anthropoid locomotor morphology (including an increase in thumb opposability in catarrhines) was initiated by selection pressure for reaching and then efficiently collecting and manipulating fruit, new leaves, buds and flowers in the small-branch, peripheral region of trees.

This, it could be said, is merely futher plausible speculation, but by placing the changes in their ecological context, e.g. size increases becoming neces-

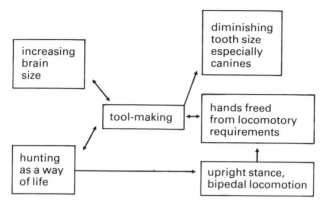

Figure 11.1 A widely postulated view of some of the interactions important in hominid evolution.

sary because of a cooler more variable climate, diurnality evolving because of size changes, as would dietary changes, both in response to changes in climate and size, Cachel is able to set up climate-related hypotheses. It may be possible to discriminate between Cachel's hypotheses and those in Figure 11.1. In the case of hypotheses relating to more recent micro-evolution, it may be possible to test them more precisely where historical evidence is also available (cf. e.g. Piazza *et al.*, 1981) and they thus become less "fossil-proof", as Pilbeam (1980) described early attempts at theories of hominid evolution.

When one understands that the "story" element of an explanation is not an end in itself (cf. Freud, 1930, for examples of the converse), much of the criticism becomes irrelevant, though even here the question of which of a multiplicity of choices is correct will arise. As Pilbeam (1980) pointed out, the current adaptive significance of some trait or structure may not relate to its original adaptive role. Bipedalism, for instance, is ordinarily regarded as having arisen, as is implied in the examples above, and having become important in the context of hunting or perhaps herbivory, yet today some vegetarian baboons adopt bipedal attitudes more frequently when ground feeding (Rose, 1976), though not adapted to bipedality as such (Lovejoy, 1981).

There remain two additional, important, related problems. The first is that natural selection, to the extent that it has been observed and measured, has been a short-term process. This concerns some people. For example, Platnick (1979) wrote "all selection in the world hasn't changed the fact that what were two morphs of *Biston betularia* are still (only) two morphs of *Biston betularia*". Platnick's point of view, and it is not an uncommon one, is that of God in Isaac Watts' quatrain:

A thousand ages in Thy sight
  Are like an evening gone;
Short as the watch that ends the night
  Before the rising sun

The outlook of the geneticist, theoretical or experimental, is more human and perhaps more humble.

Related to this problem is that exemplified by Kacser and Burns (1981), who wrote, in a critique of the hypothesis of evolution of dominance

> it is . . . perfectly possible that a *particular* dominance relationship has been the subject of evolutionary modification. This, however, is a historical question and hence not subject to experimental verification by reference to present observations.

Given the long time over which evolution has occurred, the extinction of most taxa that have ever existed, and the short life span of humans, this would seem to suggest that no evolutionary hypothesis was subject to experimental verification. This would be to deny the normal *modus operandi* of biological scientists, who behave as if "there are such things in nature as parallel cases; that what happens once will, under a sufficient degree of similarity of circumstance, happen again", as J. S. Mill (1879) put it. The key problem is to define the relevant circumstances and make them similar enough.

One reasonably strong conclusion may be drawn. The major constraint on natural selection as an agent of change is natural selection as a stabilizing force. Canalization is the most obvious result of this process, but stabilizing selection is important in itself. Consider again the example of human birth-weight. If a significant proportion of the disadvantage of extreme-weighted babies relates to their size, as such, rather than to other correlated traits, then this trait is perhaps associated with $0 \cdot 1$–1% of the force of mortality in a good environment. Other attributes of size and shape are almost certainly selected similarly later in life (Mitton, 1975; Lasker and Thomas, 1976; Bailey and Garn, 1979; Mueller *et al.*, 1981). It would not require many such traits to render directional selection ineffective, as has been pointed out for artificial selection (Minvielle, 1980; Nicholas and Robertson, 1980).

How does change come about? Minor random change occurs in small populations. More importantly, the environment is always changing, usually for the worse, except in rare cases, as perhaps for nematodes in soil or *Latimeria* and other inhabitants of the ocean depths (though even *Latimeria's* proteins may have changed, as have those of Limulus; see Selander *et al.*, 1970; Kimura and Ohta, 1971). A second escape from stabilizing forces is heterochrony in ontongeny (Gould, 1977; Alberch *et al.*, 1979), as discussed in Chapter 3. Thus, in these ways the stabilizing force is overcome, and when this happens, much can change very rapidly.

Changes will occur; this can be predicted with great confidence. But the direction of change, as I have tried to emphasize, is hardly predictable. When the skiing pioneer Arnold Lunn challenged J. B. S. Haldane to name an incipient organ or function which was currently evolving *de novo*, Haldane was unable to answer him satisfactorily, in the sense of citing a completely indisputable case (Lunn and Haldane, 1935). This is still a problem (cf. Henle, 1942; Edson *et al.*, 1981). However, it reflects once more our transience in an evolutionary perspective. Population genetics shows us some of the means to evolutionary change, but the historical sciences show us the ends.

TOPICS FOR DISCUSSION

1. "Knowledge in most disciplines is grouped into areas of thought that are built on the pattern of Euclidean geometry." (Lewis, 1980).
   What is required for this to be an appropriate description of the theory of natural selection?
2. "Perhaps the most interesting area of evolution – the origins of the human lineage – can only be cautiously dealt with due to incomplete evidence. It may be the most likely period of time when rapid or explosive evolution may have occurred. The earliest undoubted hominids in the fossil record . . . are quite primitive and date to somewhat less than 4·0 Myr. The molecular clock suggests a hominid–pongid divergence somewhat before this . . . . Thus, the hominid transition could have been characterized by extremely rapid morphological change (the evolution of bipedality), and by a period of low population numbers, or bottlenecks, with most of chromosomal evolution, genic reorganization, or allelic substitutions occurring in rapid sequence. It is here that the evidence fails us in being particularly incomplete, only partially suggestive and open to many different interpretations . . . , and it may be that as more evidence becomes available even this possible case of punctuation may be like the cheshire cat and disappear until only its smile is left." (Cronin *et al.*, 1981).
   What evidence would be necessary to show that abrupt changes, at the rates discussed in Chapter 6 for traits like melanism in certain moths, are qualitatively distinct from relatively rapid changes like those in hominid evolution?
3. "Out of the darkness of prehistoric ages man emerges with the marks of his lowly origin strong upon him. He is a brute, only more intelligent than the other brutes, and blind prey to impulses, which as often as not lead him to destruction; a victim to endless illusions, which make his mental existence a terror and a burden, and fill his life with barren toil and battle." (Huxley, 1889).
   What implications has the theory of evolution by natural selection for this view of man?

references

Abbie, A. A. (1939). The origin of the corpus callosum and the fate of the structures related to it. *J. Comp. Neur.* **70**, 9–44.

Abbie, A. A. (1941). Marsupials and the evolution of mammals. *Aust. J. Sci.* **4**, 77–92.

Ahearn, J. N. (1980). Evolution of behavioural reproductive isolation in a laboratory stock of *Drosophila silvestris*. *Experientia* **36**, 63–4.

Alberch, P., Gould, S. J., Oster, G. F. and Wake, D. B. (1979). Size and shape in ontogeny and phylogeny. *Paleobiology* **5**, 296–317.

Alexander, R. D. (1974). The evolution of social behaviours. *Ann. Rev. Ecol. Syst.* **5**, 325–83.

Allendorf, F. W. (1979). Rapid loss of duplicate gene expression by natural selection. *Heredity* **43**, 247–58.

Allison, A. C. (1955). Aspects of polymorphism in man. *Cold Spring Harb. Symp. Quant. Biol.* **20**, 239–55.

Alvarez, L. W., Alvarez, W., Asaro, F. and Michel, H. V. (1980). Extraterrestrial cause for the Cretaceous–Tertiary extinction. *Science* **208**, 1095–108.

Anderson, E. (1948). Hybridization of the habitat. *Evolution* **2**, 1–9.

Anderson, E. and Stebbins, G. L. Jr (1954). Hybridization as an evolutionary stimulus. *Evolution* **8**, 378–88. In *Hybridization An Evolutionary Perspective* (ed. D. A. Levin). Dowden, Hutchinson and Ross Inc.

Apple, M. S. and Korostyshevskiy, M. A. (1980). Why many biological parameters are connected by power dependence. *J. Theor. Biol.* **85**, 569–73.

Arbous, A. G. and Kerrich, J. E. (1951). Accident statistics and the concept of accident-proneness. *Biometrics* **1**, 340–52.

Archibald, J. D. (1981). The earliest known Palaeocene mammal fauna and its implications for the Cretaceous–Tertiary transition. *Nature, Lond.* **291**, 650–2.

Armstrong, R. A. and McGehee, R. (1980). Competitive exclusion. *Am. Nat.* **115**, 151–70.

Arnheim, N., Krystal, M., Schmickel, R., Wilson, G., Ryder, O. and Zimmer, E. (1980). Molecular evidence for genetic exchanges among ribosomal genes on non-homologous chromosomes in man and apes. *Proc. Natl Acad. Sci., USA* **77**, 7323–7.

Astolfi, P., Kidd, K. K. and Cavalli-Sforza, L. L. (1981). A comparison of methods for reconstructing evolutionary trees. *Syst. Zool.* **30**, 156–69.

Atchley, W. R. and Rutledge, J. J. (1980). Genetic components of size and shape. I. Dynamics of components of phenotypic variability and covariability during ontogeny in the laboratory rat. *Evolution* **34**, 1161–73.

Avise, J. C., Patton, J. C. and Aquadro, C. F. (1980a). Evolutionary genetics of birds. Comparative molecular evolution in new world warblers and rodents. *J. Hered.* **71**, 303–10.

Avise, J. C., Patton, J. C. and Aquadro, C. F. (1980b). Evolutionary genetics of birds. II. Conservative protein evolution in North American sparrows and relatives. *Syst. Zool.* **89**, 323–34.

Axelrod, R. and Hamilton, W. D. (1981). The evolution of co-operation. *Science* **211**, 1390–6.

Baba, M. L., Darga, L. L., Goodman, M. and Czelusniak, J. (1981). Evolution of cytochrome *c* investigated by the maximum parsimony method. *J. Mol. Evol.* **17**, 197–213.

Bader, R. S. (1955). Variability and evolutionary rate in the oreodonts. *Evolution* **9**, 119–40.

Bailey, S. M. and Garn, S. M. (1979). Socio-economic interactions with physique and fertility. *Hum. Biol.* **51**, 317.

Bailey, G. S., Poulter, R. T. M. and Stockwell, P. A. (1978). Gene duplication in tetraploid fish: model for gene silencing at unlinked duplicate loci. *Proc. Natl Acad. Sci., USA* **75**, 5575–9.

Baker, R. J. and Bickham, J. W. (1980). Karyotypic evolution in bats: evidence of extensive and conservative chromosomal evolution in closely related taxa. *Syst. Zool.* **29**, 239–53.

Baldwin, J. E. and Krebs, H. (1981). The evolution of metabolic cycles. *Nature Lond.* **291**, 381–82.

Ball, J. A. (1980). Extraterrestrial intelligence: where is everybody? *Am. Sci.* **68**, 656–63.

Barbieri, M. (1981). The ribotype theory of the origin of life. *J. Theor. Biol.* **91**, 545–61.

Barlow, H. B. (1981). Critical limiting factors in the design of the eye and visual cortex. *Proc. Roy. Soc. Lond., B* **212**, 1–34.

Bateman, A. J. (1952). Self-incompatibility in angiosperms. I. Theory. *Heredity* **6**, 285–310.

Beach, J. H. and Bawa, K. S. (1980). Role of pollinators in the evolution of dioecy from distyly. *Evolution* **34**, 1138–42.

Beckman, G. (1978). Enzyme polymorphism. In *The Biochemical Genetics of Man* (eds D. J. H. Brock and O. Mayo). Academic Press, London and New York.

Beckner, M. (1959). *The Biological Way of Thought.* Columbia University Press, New York.

Bell, G. (1980). The costs of reproduction and their consequences. *Am. Nat.* **116**, 45–76.

Bellingham, A. J. (1976). Haemoglobins with altered oxygen affinity. *Br. Med. Bull.* **32**, 234–8.

Bengtsson, B. O. (1980a). Rates of karyotype evolution in placental mammals. *Hereditas* **92**, 37–47.

Bengtsson, B. O. (1980b). How evolutionarily stable is the XX/XY sex-determination system? *Adv. Appl. Prob.* **12**, 1–13.

Bengtsson, B. O. and Bodmer, W. F. (1976a). On the increase of chromosome mutations under random mating. *Theor. Pop. Biol.* **9**, 260–81.

Bengtsson, B. O. and Bodmer, W. F. (1976b). The fitness of human translocation carriers. *Ann. Hum. Genet.* **40**, 253–7.

Bennett, J. H. (1954). On the theory of random mating. *Ann. Eugen.* **18**, 311–17.

Bennett, J. H. (1976). Expectations for inbreeding depression on self-fertilization of tetraploids. *Biometrics* **32**, 449–52.

Bennett, M. D. (1972). Nuclear DNA content and minimum generation time in herbaceous plants. *Proc. Roy. Soc. Lond., B.* **181**, 109–35.

Bergson, H. (1907). *L'Évolution Créatrice.* Paris.

Bickham, J. W. (1981). Two-hundred-million-year-old chromosomes: deceleration of the rate of karyotypic evolution in turtles. *Science* **212**, 1291–3.

Bickham, J. W. and Baker, R. J. (1979). Canalization model of chromosomal evolution. *Bull. Carnegie Mus. Nat. Hist.* No. 13.

Bishop, J. A. (1980). Commentary on three papers on melanism. *Proc. Roy. Soc. Lond., B* **210**, 273–5.

Blaustein, A. R. and O'Hara, R. K. (1981). Genetic control for sibling recognition? *Nature, Lond.* **290**, 246–8.

Blundell, R., Lindley, P., Miller, L., Moss, D., Slingsby, C., Tickle, I., Turnell, B. and Wistow, G. (1981). The molecular structure and stability of the eye lens: X-ray analysis of γ-crystallin II. *Nature, Lond.* **289**, 771–7.

Bookchin, R. M., Nagel, R. L. and Ranney, H. M. (1970). The effect of β^{73Asn} on the interactions of sickling hemoglobins. *Biochem. Biophys. Acta* **221**, 373–5.

Bookstein, F. L., Gingerich, P. D. and Kluge, A. G. (1978). Hierarchical linear modeling of the tempo and mode of evolution. *Paleobiology* **4**, 120–34.

Boucot, A. J. (1976). Rates of size increase and of phyletic evolution. *Nature, Lond.* **261**, 695–6.

Boyd, R. and Richerson, P. J. (1980). Effect of phenotypic variation on kin selection. *Proc. Natl Acad. Sci., USA* **77**, 7506–9.

Boyer, S. H., Scott, A. F., Kunkel, L. M. and Smith, K. D. (1978). The proportion of all point mutations which are unacceptable: an estimate based on haemoglobin amino acid and nucleotide sequences. *Can. J. Genet. Cytol.* **20**, 111–37.

Bradshaw, A. D., McNeilly, T. S. and Gregory, R. P. G. (1965). Industrialisation, evolution and the development of heavy metal tolerances in plants. *5th Symp. Bri. Ecol. Soc.,* pp. 327–43. Blackwell, Oxford.

Brady, R. H. (1979). Natural selection and the criteria by which a theory is judged. *Syst. Zool.* **28**, 600–21.

Bramwell, C. D. and Whitfield, G. R. (1974). Biomechanics of Pteranodon. *Phil. Trans. Roy. Soc. Lond., B* **267**, 503–81.

Brandham, P. E. (1977). The frequency of spontaneous structural change. In *Current Chromosome Research* (eds K. Jones and P. E. Brandham). Elsevier, North Holland.

Brent, L., Rayfield, L. S., Chandler, P., Fierz, W., Medawar, P. B. and Simpson, E. (1981). Supposed Lamarckian inheritance of immunological tolerance. *Nature, Lond.* **290**, 508–12.

Bridges, C. B. (1935). Salivary gland chromsome maps. *J. Hered.* **26**, 60–4.

Britten, R. J. and Davidson, E. H. (1969). Gene regulation for higher organisms: a theory. *Science* **165**, 349–57.

Brown, W. L. Jr and Wilson, E. O. (1956). Character displacement. *Syst. Zool.* **5**, 49–64.

Bull, C. M. (1979). A narrow hybrid zone between two Western Australian frog species *Ranidella insignifera* and *R. pseudinsignifera*: the fitness of hybrids. *Heredity* **42**, 381–9.

Burley, N. (1980). Clutch overlap and clutch size: alternative and complementary reproductive tactics. *Am. Nat.* **115**, 223–46.

Bush, G. L. (1975). Modes of animal speciation. *Ann. Rev. Ecol. Syst.* **6**, 339–64.

Bush, G. L., Case, S. M., Wilson, A. C. and Patton, J. L. (1977). Rapid speciation and chromosomal evolution in mammals. *Proc. Natl Acad. Sci., USA* **74**, 3942–6.

Buss, L. W. (1981). Group living, competition, and the evolution of cooperation in a sessile invertebrate. *Science* **213**, 1012–14.

Cachel, S. (1979). A paleoecological model for the origin of higher primates. *J. Hum. Evol.* **8**, 351–9.

Calow, P. (1979). The cost of reproduction – a physiological approach. *Biol. Rev.* **54**, 23–40.

Campbell, R. B. (1981). Some circumstances assuring monomorphism. *Theor. Pop. Biol.* **20**, 118–25.

Carlson, S. S., Wilson, A. C. and Maxson, R. D. (1978). Do albumin clocks run on time? *Science* **200**, 1183–5.

Carr, B. J. and Rees, M. J. (1979). The anthropic principle and the structure of the physical world. *Nature, Lond.* **278**, 605–12.

Carson, H. L. (1975). The genetics of speciation at the diploid level. *Am. Nat.* **109**, 83–92.

Carson, H. L. and Bryant, P. J. (1979). Change in a secondary sexual character as evidence of incipient speciation in *Drosophila silvestris*. *Proc. Natl Acad. Sci., USA* **76**, 1929–32.

Case, T. J. and Bender, E. A. (1981). Is recombination advantageous in fluctuating and spatially heterogeneous environments? *J. Theor. Biol.* **90**, 181–90.

Cavalli-Sforza, L. L. and Feldman, M. W. (1978). Darwinian evolution and "altruism". *Theor. Pop. Biol.* **14**, 268–80.

Cavener, D. R. and Clegg, M. T. (1981). Evidence for biochemical and physiological differences between enzyme genotypes in *Drosophila melanogaster*. *Proc. Natl Acad. Sci., USA* **78**, 4444–7.

Cedergren, R. J., Larue, B., Sankoff, D., Lapalme, G. and Grosjean, J. (1980). Convergence and minimal mutation criteria for evaluating early events in tRNA evolution. *Proc. Natl Acad. Sci., USA* **77**, 2791–5.

Chakraborty, R. (1981). Estimation of mutation rates from the number of rare alleles in a sample. *Ann. Hum. Biol.* **8**, 221–30.

Chapman, R. W. (1981). Recombination potentials and evolutionary rates. *Am. Nat.* **118**, 384–93.

Charlesworth, B. (1979). Selection for gamete lethals and S-alleles in complex heterozygotes. *Heredity* **43**, 159–64.

Charlesworth, B. (1980). The cost of sex in relation to mating system. *J. theor. Biol.* **84**, 655–71.

Charlesworth, B. and Charlesworth, D. (1973). Selection of new inversions in multi-locus genetic systems. *Genet. Res.* **21**, 167–83.

Charlesworth, D. and Charlesworth, B. (1979). The evolution and breakdown of S-allele systems. *Heredity* **43**, 41–55.

Charnov, E. L. and Krebs, J. H. (1974). On clutch-size and fitness. *Ibis* **116**, 217–19.

Chomsky, N. (1976). On the nature of language. *Ann. N. Y. Acad. Sci.* **280**, 46–55.

Cifelli, R. A. (1981). Patterns of evolution among the artiodactyla and perissodactyla (Mammalia). *Evolution* **35**, 433–40.

Clarke, C. A. and Sheppard, P. M. (1963). Interactions between major genes and polygenes in the determination of the mimetic patterns of *Papilio dardanus*. *Evolution* **17**, 404–13.

Clarke, C. A. and Sheppard, P. M. (1970). Is *Papilio gothica* (Papilionidae) a good species? *J. Lepidopterists' Soc.* **24**, 229–33.

Clarke, C. A. and Sheppard, P. M. (1971). Further studies on the genetics of the mimetic butterfly *Papilio memnon* L. *Phil. Trans. Roy. Soc. Lond.*, *B* **263**, 35–70.

Clarke, C. A. and Sheppard, P. M. (1972). The genetics of the mimetic butterfly *Papilio polytes* L. *Phil. Trans. Roy. Soc. Lond., B.* **263**, 431–58.

Clarke, C. A. and Sheppard, P. M. (1973). The genetics of four new forms of the mimetic butterfly *Papilio memnon* L. *Proc. Roy. Soc. Lond., B.* **184**, 1–14.

Clarke, C. A. and Sheppard, P. M. (1977). A new tailed female form of *Papilio memnon* L. and its probable genetic control. *Syst. Entomol.* **2**, 17–19.

Clarke, C. A., Sheppard, P. M. and Thornton, I. W. B. (1968). The genetics of the mimetic butterfly *Papilio memnon. Phil. Trans. Roy. Soc. Lond., B.* **254**, 37–89.

Clarke, P. H. (1980). Experiments in microbial evolution: new enzymes, new metabolic activities. *Proc. Roy. Soc. Lond., B* **207**, 385–404.

Clemens, W. and Archibald, D. (1980). Evolution of terrestrial faunas during the Cretaceous–Tertiary transition. *Mém. Soc. géol. Fr., N.S.* **139**, 67–74.

Cloud, P. and Morrison, K. (1979). On microbial contaminants, micropseudofossils, and the oldest records of life. *Precambrian Res.* **9**, 81–91.

Cloud, P., Gustafson, L. B. and Watson, J. A. L. (1980). The works of living social insects as pseudofossils and the age of the oldest known Metazoa. *Science* **210**, 1013–15.

Clutton-Brock, T. H. (1980). Primates, brains and ecology. *J. Zool., Lond.* **190**, 309–23.

Clutton-Brock, T. H. and Harvey, P. H. (1979). Home range size, population density and phylogeny in primates. In *Primate Ecology and Human Origins* (eds I. S. Bernstein and E. O. Smith). pp. 201–14. Garland Press, New York.

Clutton-Brock, T. H. and Harvey, P. H. (1980). Primates, brains and ecology. *J. Zool., Lond.* **190**, 309–23.

Cody, M. L. (1966). A general theory of clutch size. *Evolution* **20**, 174–84.

Colgan, D. J. (1981). The relative viabilities of alternative human genotypes. *Human Heredity* **31**, 172–9.

Colwell, R. K. (1981). Group selection is implicated in the evolution of female-biassed sex ratios. *Nature, Lond.* **290**, 401–4.

Connor, E. F. and Simberloff, D. (1979). The assembly of species communities: chance or competition? *Ecology* **60**, 1132–40.

Cooper, D. W., Johnston, P. G., Vandeberg, J. L., Maynes, G. M. and Chew, G. K. (1979). A comparison of genetic variability at X-linked and autosomal loci in kangaroos, man and *Drosophila. Genet. Res.* **33**, 243–52.

Corbin, K. W. and Uzzell, T. (1970). Natural selection and mutation rates in mammals. *Am. Nat.* **104**, 37–53.

Cracraft, J. (1974). Phylogenetic models and classification. *Syst. Zool.* **23**, 71–91.

Cronin, J. E., Boaz, N. T., Stringer, C. B. and Rak, Y. (1981). Tempo and mode in hominid evolution. *Nature, Lond.* **292**, 113–22.

Crosby, J. L. (1940). High proportion of homostyle plants in populations of *Primula vulgaris. Nature, Lond.* **145**, 672.

Crow, J. F. (1958). Some possibilities for measuring selection intensities in man. *Hum. Biol.* **30**, 1–13.

Crow, J. F. (1961). Mutation in man. *Progr. med. Genet.* **1**, 1–26.

Crow. J. F. (1966). The quality of people: human evolutionary changes. *BioScience* **16**, 863–7.

Crow, J. F. and Nagylaki, T. (1976). The rates of change of a character correlated with fitness. *Am. Nat.* **110**, 207–13.

Cruden, R. W. and Hermann-Parker, S. M. (1977). Temporal dioecism: an alternative to dioecism? *Evolution* **31**, 863–6.

Cuénot, L. (1914). Cited by Fisher and Stock (1915), q. v.

Cuénot, L. (1951). *L'Évolution Biologique*. Masson et Cie, Paris.

Cullis, C. A. (1977). Molecular aspects of the environmental induction of heritable changes in flax. *Heredity* **38**, 129–54.

Darlington, C. D. (1939). *The Evolution of Genetic Systems*. Cambridge University Press, Cambridge and London.

Darlington, C. D. (1956). *Chromosome Botany*. George Allen and Unwin, London.

Darlington, P. J. (1981). Genes, individuals and kin selection. *Proc. Natl Acad. Sci. USA* **78**, 4440–43.

Darwin, C. R. (1859). *On the Origin of Species by Means of Natural Selection, or the Preservation of Favoured Races in the Struggle for Life*. John Murray, London.

Darwin, C. R. (1868). *Variation of Animals and Plants under Domestication*. John Murray, London.

Darwin, C. R. (1871). *The Descent of Man, and Selection in Relation to Sex*. John Murray, London.

Darwin, C. R. (1882). Published as *The autobiography of Charles Darwin 1809–1882* with original omissions restored. Edited with appendix and notes by his granddaughter Nora Barlow. Collins, London.

Davenport, C. B. (1909). Mutation. In *Fifty years of Darwinism*, pp. 160–181. Henry Holt, New York.

Dawkins, R. (1979). Twelve misunderstandings of kin selection. *Z. Tierpsychol.* **51**, 184–200.

De Jong, W. W. and Rydén, L. (1981). Causes of more frequent deletions than insertions in mutations and protein evolution. *Nature, Lond.* **290**, 157–9.

De Jong, W. W., Greaves, J. T. and Boulter, D. (1977). Evolutionary changes of α-crystallin and the phylogeny of mammalian orders. *J. Mol. Evol.* **10**, 123–35.

Demetrius, L. (1974). Demographic parameters and natural selection. *Proc. Natl Acad. Sci., USA*, **71**, 4645–7.

de Vries, H. (1902). *Die mutationstheorie*. Berlin.

de Wet, J. M. J. (1979). Origins of polyploidy. In *Polyploidy Biological Relevance* (ed. W. H. Lewis), pp. 3–15. Plenum, New York.

Diamond, J. M. and Veitch, C. R. (1981). Extinction and introductions in the New Zealand avifauna: cause and effect? *Science* **211**, 499–511.

Dickinson, H. and Antonovics, J. (1973). Theoretical considerations of sympatric divergence. *Am. Nat.* **107**, 256–74.

Dirac, P. A. M. (1937). The cosmological constraints. *Nature, Lond.* **139**, 323.

Dobzhansky, Th. (1937). *Genetics and the Origin of Species*. Columbia University Press, New York.

Dobzhansky, Th. (1970). *Genetics of the Evolutionary Process*. Columbia University Press, New York.

Dodson, M. M. (1975). Quantum evolution and the fold catastrophe. *Evol. Theor.* **1**, 107–18.

Dodson, M. M. (1976). Darwin's law of natural selection and Thom's theory of catastrophes. *Math. Biosciences* **28**, 243–74.

Dodson, M. M. and Hallam, A. (1977). Allopatric speciation and the fold catastrophe. *Am. Nat.* **111**, 415–33.

Doolittle, W. F. and Sapienza, C. (1980). Selfish genes, the phenotype paradigm and genome evolution. *Nature, Lond.* **284**, 601–3.

Duncan, C. J. and Sheppard, P. M. (1963). Continuous and quantal theories of sensory discrimination. *Proc. Roy. Soc. Lond., B* **158**, 343–63.

Duncan, C. J. and Sheppard, P. M. (1965). Sensory discrimination and its role in the evolution of Batesian mimicry. *Behaviour* **24**, 269–82.

Durrant, A. (1958). Environmental conditioning of flax. *Nature, Lond.* **181**, 928–9.

Eanes, W. F. (1978). Morphological variance and enzyme heterozygosity in the monarch butterfly. *Nature, Lond.* **276**, 263–4.

Eberhard, W. G. (1980). Evolutionary consequences of intracellular organelle competition. *Quart. Rev. Biol.* **55**, 231–49.

Economos, A. C. (1981). The largest land mammal. *J. Theor. Biol.* **89**, 211–15.

Edson, M. M., Foin, T. C. and Knapp, C. M. (1981). "Emergent properties" and ecological research. *Am. Nat.* **118**, 593–6.

Edwards, J. H. (1972). Advantageous detrimentals and the neutral gene. *Nature New Biol.* **236**, 250.

Edwardson, J. R. (1970). Cytoplasmic male sterility. *Bot. Rev.* **36**, 341–420.

Eigen, M. (1971). Self organization of matter and the evolution of biological macromolecules. *Naturwissenschaften* **58**, 465–523.

Eigen, M. and Schuster, P. (1979). *The Hypercycle: A Principle of Natural Self-organization.* Springer, Berlin.

Eigen, M., Gardiner, W. C. Jr and Schuster, P. (1980). Hypercycles and compartments. *J. Theor. Biol.* **85**, 407–11.

Eigenmann, C. H. (1909). Adaptation. In *Fifty years of Darwinism* pp. 182–208. Henry Holt, New York.

Eldredge, N. and Gould, S. J. (1972). Punctuated equilibria: An alternative to phyletic gradualism. In *Models in Paleobiology* (ed. T. J. M. Schopf), pp. 82–115. Freeman, Cooper and Co., San Francisco.

Ewens, W. J. (1979a). Testing the generalised neutrality hypothesis. *Theor. Pop. Biol.* **15**, 205–16.

Ewens, W. J. (1979b). *Mathematical Population Genetics.* Berlin, Springer-Verlag.

Ewens, W. J. and Thomson, G. (1977). Properties of equilibria in multi-locus genetic systems. *Genetics* **87**, 807–19.

Falconer, D. S. (1960). *Introduction to Quantitative Genetics.* Oliver and Boyd, Edinburgh.

Falconer, D. S. (1966). Genetic consequences of selection pressure. *Genetic and Environmental Factors in Human Ability* (eds J. E. Meade and A. S. Parkes) pp. 219–32. Oliver and Boyd, Edinburgh.

Farris, J. S. (1966). Estimation of conservatism by constancy within biological populations. *Evolution* **20**, 587–91.

Farris, J. S. (1970). On the relationship between variation and conservatism. *Evolution* **24**, 825–7.

Feldman, M. (1976). Wheats. In *Evolution of Crop Plants* (ed. N. W. Simmonds) pp. 120–8. Longman, London.

Feldman, M. W. and Cavalli-Sforza, L. L. (1981). Further remarks on Darwinian selection and "altruism". *Theor. Pop. Biol.* **19**, 251–60.

Feldman, M. W., Christiansen, F. B. and Brooks, L. D. (1980). Evolution of recombination in a constant environment. *Proc. Natl Acad. Sci., USA* **77**, 4838–41.

Felsenstein, J. (1974). The evolutionary advantage of recombination. *Genetics* **78**, 737–56.

Felsenstein, J. (1978a). Macroevolution in a model ecosystem. *Am. Nat.* **112**, 177–95.

Felsenstein, J. (1978b). The number of evolutionary trees. *Syst. Zool.* **27**, 27–33.

Felsenstein, J. (1979). *r*- and *K*-selection in a completely chaotic population model. *Am. Nat.* **113**, 499–510.

Felsenstein, J. (1981). Skepticism towards Santa Rosalia, or why are there so few kinds of animals? *Evolution* **35**, 124–38.

Felsenstein, J. and Yokoyama, S. (1976). The evolutionary advantage of recombination. II. Individual selection for recombination. *Genetics* **83**, 845–59.

Ferris, S. D., Wilson, A. C. and Brown, W. M. (1981). Evolutionary tree for apes and humans based on cleavage maps of mitochondrial DNA. *Proc. Natl Acad. Sci., USA* **78**, 2432–6.

Fisher, R. A. (1914). Some hopes of a eugenist. *Eugen. Rev.* **5** 309–15.

Fisher, R. A. (1918). The correlation between relatives on the supposition of Mendelian inheritance. *Trans. Roy. Soc. Edin.* **52**, 399–433.

Fisher, R. A. (1928*a*). The possible modification of the response of the wild type to recurrent mutations. *Am. Nat.* **62**, 115–26.

Fisher, R. A. (1928*b*). Two further notes on the origin of dominance. *Am. Nat.* **62**, 571–4.

Fisher, R. A. (1929). The evolution of dominance; reply to Professor Sewell Wright. *Am. Nat.* **63**, 553–6.

Fisher, R. A. (1930*a*). The evolution of dominance in certain polymorphic species. *Am. Nat.* **64**, 385–406.

Fisher, R. A. (1930*b*). *The Genetical Theory of Natural Selection.* Clarendon Press, Oxford. (Second edition, Dover Publications, New York, 1958.)

Fisher, R. A. (1931). The evolution of dominance. *Biol. Rev.* **6**, 345–68.

Fisher, R. A. (1933*a*). Selection in the production of the eversporting stocks. *Ann. Bot.* **47**, 727–33.

Fisher, R. A. (1933*b*). On the evidence against the chemical induction of melanism in Lepidoptera. *Proc. Roy. Soc. Lond., B* **112**, 407–11.

Fisher, R. A. (1936). "The coefficient of racial likeness" and the future of craniometry. *J. Roy. Anthropological Inst.* **66**, 57–63.

Fisher, R. A. (1939). Selective forces in wild populations of *Paratettix texanus. Ann. Eugen.* **9**, 109–22.

Fisher, R. A. (1941). Average excess and average effect of a gene substitution. *Ann. Eugen.* **11**, 53–63.

Fisher, R. A. (1949). A theoretical system of selection for homostyle *Primula. Sankhyā* **9**, 325–42.

Fisher, R. A. (1954). Retrospect of the criticisms of the theory of natural selection. In *Evolution as a Process* (eds J. S. Huxley, A. C. Hardy and E. B. Ford) pp. 84–98. Allen and Unwin, London.

Fisher, R. A. (1958). The discontinuous inheritance. *Listener* **60**, 85–7.

Fisher, R. A. and Ford, E. B. (1947). The spread of a gene in natural conditions in a colony of the moth *Panaxia dominula* L. *Heredity* **1**, 143–74.

Fisher, R. A. and Gray, H. (1937). Inheritance in Man: Boas's data studied by the method of analysis of variance. *Ann. Eugen.* **8**, 74–93.

Fisher, R. A. and Stock, C. S. (1915). Cuénot on preadaptation. A criticism. *Eugen. Rev.* **7**, 46–61.

Fitch, W. M. (1976). Is there selection against wobble in codon–anticodon pairing? *Science* **194**, 1173–4.

Ford, E. B. (1975). *Ecological Genetics*, 5th edn. Chapman and Hall, London.

Foster, G. G., Whitten, M. J., Konovalov, C., Arnold, J. T. A. and Maffi, G. (1981). Autosomal genetic maps of the Australian Sheep Blowfly, *Lucilia cuprina dorsalis*

R.-D. (Diptera: Calliphoridae), and possible correlations with the linkage maps of *Musca domestica* L. and *Drosophila melanogaster* (Mg.). *Genet. Res.* **37**, 55–69.

Fox, G. W., Stackebrandt, E., Hespell, R. B., Gibson, J., Maniloff, J., Dyer, T. A., Wolfe, R. S., Balch, W. E., Tanner, R. S., Magrum, L. J., Zablen, L. B., Blakemore, R., Gupta, R., Bonen, L., Lewis, B. J., Stahl, D. A., Luehrsen, K. R., Chen, K. N. and Woese, C. R. (1980). The phylogeny of prokaryotes. *Science* **209**, 457–63.

Fraccaro, M. (1956). A contribution to the study of birth weight based on an Italian sample. *Ann. Hum. Genet.* **20**, 282–98.

Francke, U. and Taggart, R. T. (1980). Comparative gene mapping: order of loci on the X chromosome is different in mice and humans. *Proc. Natl Acad. Sci., USA* **77**, 3595–9.

Franklin, I. and Lewontin, R. C. (1970). Is the gene the unit of selection? *Genetics* **65**, 707–34.

Freeman, D. D., Klikoff, L. G. and Harper, K. T. (1976). Differential resource utilization by the sexes of dioecious plants. *Science* **193**, 597–9.

Freire-Maia, N. and Robson, E. B. (1979). Notes: neutralist hypothesis is Darwinian. *Ann. Hum. Genet.* **42**, 531.

Freud, S. (1930). *Civilization and its Discontents*. Hogarth Press, London.

Ganapathy, R. (1980). A major meteorite impact on the earth 65 million years ago: from the Cretaceous–Tertiary boundary clay. *Science* **209**, 921–3.

Gans, C. (1979). Momentarily excessive construction as the basis for protoadaptation. *Evolution* **33**, 227–33.

Gartner, S. and Keany, J. (1978). The terminal Cretaceous event: a geological problem with an oceanographic solution. *Geology* **6**, 708–12.

Gartner, S. and McGuirk, J. P. (1979). Terminal Cretaceous extinction: scenario for a catastrophe. *Science* **206**, 1272–6.

Georgeson, M. A. and Harris, M. G. (1981). Size constancy does not fail below half a degree. *Nature, Lond.* **289**, 826.

Gibbon, E. (1776). *The Decline and Fall of the Roman Empire*, Vol. 1. London.

Gillespie, J. H. (1974). Natural selection for within-generation variance in offspring number. *Genetics* **76**, 601–6.

Gillespie, J. H. (1976). A general model to account for enzyme variation in natural populations. II. Characterization of the fitness function. *Am. Nat.* **110**, 809–21.

Gillespie, J. H. (1977*a*). A general model to account for enzyme variation in natural populations. IV. The quantitative genetics of viability mutants. *Measuring Selection in Natural Populations* (ed. F. B. Christiansen and T. M. Fenchel), pp. 301–14. Springer-Verlag, Berlin.

Gillespie, J. H. (1977*b*). Natural selection for variance in offspring numbers: a new evolutionary principle. *Am. Nat.* **111**, 1010–14.

Gillespie, J. H. (1978). A general model to account for enzyme variation in natural populations. V. The SAS–CFF model. *Theor. pop. Biol.* **14**, 1–45.

Gingerich, P. D. (1980). Evolutionary patterns in early cenozoic mammals. *Ann. Rev. Earth Planet Sci.* **8**, 407–24.

Gingerich, P. D. and Simons, E. L. (1977). Systematics, phylogeny, and evolution of early Eocene Adapidae (Mammalia, Primates) in North America. *Contrib. Mus. Paleontol. Univ. Mich.* **24**, 245–79.

Gladyshev, G. P. and Khasanov, M. M. (1981). Optical activity and evolution. *J. theor. Biol.* **90**, 191–8.

Godfrey, L. and Jacobs, K. H. (1981). Gradual, autocatalytic and punctuational models of hominid brain evolution: a cautionary tale. *J. Hum. Evol.* **10**, 255–72.

Gold, J. R. (1980). Chromosomal change and rectangular evolution in North American cyprinid fishes. *Genet. Res.* **35**, 157–64.

Goldschmidt, R. (1940). *The Material Basis for Evolution.* Yale University Press, New Haven.

Goodman, M. (1981). Globin evolution was apparently very rapid in early vertebrates: a reasonable case against the rate-constancy hypothesis. *J. Mol. Evol.* **17**, 114–20.

Goodman, M., Moore, G. W. and Matsuda, G. (1975). Darwinian evolution in the genealogy of haemoglobin. *Nature, Lond.* **253**, 603–8.

Gorczynski, R. M. and Steele, E. J. (1980). Inheritance of acquired immunological tolerance to foreign histocompatibilty antigens in mice. *Proc. Natl Acad. Sci., USA* **77**, 2871–5.

Gorczynski, R. M. and Steele, E. J. (1981). Simultaneous yet independent inheritance of somatically acquired tolerance to two distinct H-2 antigenic haplotype determinants in mice. *Nature, Lond.* **289**, 678–81.

Gosse, E. (1907). *Father and Son.* London.

Gosse, P. (1858). *Omphalos.* London.

Gould, S. J. (1977). Eternal metaphors of palaeontology. In *Patterns of Evolution as Illustrated by the Fossil Record* (ed. A. Hallam), pp. 1–26. Elsevier Scientific, Amsterdam.

Gould, S. J. (1980). *The Panda's Thumb.* W. W. Norton, New York.

Gould, S. J. and Eldredge, N. (1977). Punctuated equilibria: the tempo and mode of evolution reconsidered. *Paleobiology* **3**, 115–51.

Gould, S. J., Raup, D. M., Sepkoski, J. J. Jr, Schopf, T. J. M. and Simberloff, D. S. (1977). The shape of evolution: a comparison of real and random clades. *Paleobiology* **3**, 23–40.

Grant, P. R. (1975). The classical case of character displacement. *Evolutionary Biology* (eds Th. Dobzhansky, M. K. Hecht and W. C. Steere) Vol. 8, pp. 237–337. Plenum Press, New York.

Green, R. F. (1980). Bayesian birds: a simple example of Oaten's stochastic model of optimal foraging. *Theor. Pop. Biol.* **18**, 255–65.

Greenewalt, C. H. (1975). The flight of birds. *Trans. Am. Phil. Soc.* **65**, 1–67.

Grey, D. R. (1980). Minimisation of extinction probabilities in reproducing populations. *Theor. Pop. Biol.* **18**, 430–43.

Griffing, J. B. (1981*a*). A theory of natural selection incorporating interaction among individuals. I. The modelling process. *J. Theor. Biol.* **89**, 635–58.

Griffing, J. B. (1981*b*). A theory of natural selection incorporating interaction among individuals. II. Use of related groups. *J. Theor. Biol.* **89**, 659–77.

Griffing, J. B. (1981*c*). A theory of natural selection incorporating interaction among individuals. III. Use of random groups of inbred individuals. *J. Theor. Biol.* **89**, 679–90.

Griffing, J. B. (1981*d*). A theory of natural selection incorporating interaction among individuals. IV. Use of related groups of inbred individuals. *J. Theor. Biol.* **89**, 691–710.

Griffiths, R. C. (1981). The number of heterozygous loci between two randomly chosen completely linked sequences of loci in two subdivided population models. *J. Math. Biol.* **12**, 251–61.

Grosberg, R. K. (1981). Competitive ability influences habitat choice in marine invertebrates. *Nature, Lond.* **290**, 700–2.

Grünberg, H. (1980). On pseudo-polymorphism. *Proc. Roy. Soc. Lond.,* B **210**, 533–48.

Gurwitsch, A. (1915). On practical vitalism. *Am. Nat.* **49**, 763–70.

Guthrie, R. D. (1969). Senescence as an adaptive trait. *Perspectives in Biology and Medicine* **12**, 313–24.

Haldane, J. B. S. (1924). A mathematical theory of natural and artificial selecion. *Proc. Camb. Phil. Soc.* **23**, 26–42.

Haldane, J. B. S. (1926). A mathematical theory of natural and artificial selecion. III. *Proc. Camb. Phil. Soc.* **23**, 363–72.

Haldane, J. B. S. (1928). *Possible Worlds*. Chatto and Windus, London.

Haldane, J. B. S. (1930). A note on Fisher's theory of the origin of dominance and on a correlation between dominance and linkage. *Am. Nat.* **64**, 87.

Haldane, J. B. S. (1932). *The Causes of Evolution*. Longmans Green, London.

Haldane, J. B. S. (1937). Physical science and philosophy. *Nature, Lond.* **139**, 1003.

Haldane, J. B. S. (1940). The relative importance of principal and modifying genes in determining some human diseases. *J. Genet.* **41**, 149–57.

Haldane, J. B. S. (1949). Suggestions as to quantitative measurement of rates of evolution. *Evolution* **3**, 51–6.

Haldane, J. B. S. (1954). The statics of evolution. In *Evolution as a Process* (eds J. S. Huxley, A. C. Hardy and E. B. Ford). George Allen and Unwin, London.

Haldane, J. B. S. (1956). The theory of selection for melanism in the Lepidoptera. *Proc. Roy. Soc. Lond., B.* **144**, 303–8.

Haldane, J. B. S. (1957). The cost of natural selection. *J. Genet.* **57**, 511–24.

Haldane, J. B. S. (1962). Evidence for heterosis in woodlice. *J. Genet.* **58**, 39–41.

Hall, B. G. (1980). On the evolution of new metabolic functions in diploid organisms. *Genetics* **96**, 1007–17.

Hamilton, W. D. (1963). The evolution of altruistic behaviour. *Am. Nat.* **97**, 354–6.

Hamilton, W. D. (1964a). The genetical evolution of social behaviour. I. *J. Theor. Biol.* **7**, 1–16.

Hamilton, W. D. (1964b). The genetical evolution of social behaviour. II. *J. Theor. Biol.* **7**, 17–52.

Hamilton, W. D. (1967). Extraordinary sex ratios. *Science* **156**, 477–88.

Hamilton, W. D. (1972). Altruism and related phenomena, mainly in social insects. *Ann. Rev. Ecol. Syst.* **3**, 193–232.

Hampé, A. (1959). Contribution à l'étude du développement et de la regulation des déficiences et des excédents dans la patte de l'embryon de poulet. *Archs Anat. Microsc. Morphol. Exp.* **48**, 345–478.

Hanbin, E. H. and Watson, D. M. S. (1914). On the flight of pterodactyle. *Argonaut. J.* **18**, 324–35.

Handford, P. (1980). Heterozygosity at enzyme loci and morphological variation. *Nature, Lond.* **286**, 261–2.

Hansen, O. (1981). Are the genes of universal grammar more than structural? *Hereditas* **95**, 213–18.

Hardin, G. (1960). The competitive exclusion principle. *Science* **131**, 1292–8.

Harper, C. W. Jr (1979). A Bayesian probability view of phylogenetic systematics. *Syst. Zool.* **28**, 547–53.

Harrison, R. G. (1980). Dispersal polymorphisms in insects. *Ann. Rev. Ecol. Syst.* **11**, 95–118.

Hart, G. E. and Langston, P. J. (1977). Chromosomal location and evolution of isozyme structural genes in hexaploid wheat. *Heredity* **39**, 263–77.

Hartl, D. L., Burla, H. and Jungen, H. (1980). Can statistical tests of neutrality detect selection? *Genetica* **54**, 185–9.

Harvey, P. H., Clutton-Brock, T. H. and Mace, G. M. (1980). Brain size and ecology in small mammals and primates. *Proc. Natl Acad. Sci., USA* **77**, 4387–9.

Haukioja, E. (1970). Clutch size of the Reed Bunting *Emberiza schoeniclus. Ornis Fennica* **47**, 101–35.

Hayman, D. L. and Martin, P. G. (1969). In *Mammalian Cytogenetics* (ed. K. Benirschke). Springer, Berlin.

Hayman, D. L. and Sharp, P. (1981). Verification of the structure of the complex sex chromosome system in *Lagorchestes conspicillatus* Gould (Maruspialia: Mammalia). *Chromosoma* **83**, 263–74.

Heckel, D. G. and Roughgarden, J. (1980). A species near its equilibrium size in a fluctuating environment can evolve a lower intrinsic rate of increase. *Proc. Natl Acad. Sci., USA* **77**, 7497–7500.

Hed, H. M. E. and Rasmuson, M. (1981). Cohort study of opportunity for selection in two Swedish 19th century parishes with a survey of other estimates. *Hum. Heredity* **31**, 78–83.

Hedrick, P. W. (1980). Hitchhiking: a comparison of linkage and partial selfing. *Genetics* **94**, 791–808.

Hendrickson, J. A. Jr (1981). Community-wide character displacement re-examined. *Evolution* **35**, 794–809.

Henle, P. (1942). The status of emergence. *J. Philos.* **39**, 483–93.

Hennig, W. (1966). *Phylogenetic Systematics.* University of Illinois Press, Urbana.

Herbers, J. M. (1981). Reliability theory and foraging by ants. *J. Theor. Biol.* **89**, 175–89.

Herm, D. (1965). Mikropäläontologisch-stratigraphische Untersuchungen in Kreideflysch zwischen Deva and Zumaya. *Z. dt. geol. Ges.* **115**, 277–348.

Hewitt, G. M. (1973). The integration of supernumerary chromosomes into the orthopteran genome. *Cold Spring Harbor Symp. Quant. Biol.* **38**, 183–94.

Heyde, C. C. (1978). On an explanation for the characteristic clutch size of some bird species. *Adv. Appl. Prob.* **10**, 723–5.

Heyde, C. C. and Schuh, H. J. (1978). Uniform bounding of probability generating functions and the evolution of reproduction rates in birds. *J. Appl. Prob.* **15**, 243–50.

Hickey, L. J. (1981). Land plant evidence compatible with gradual, not catastrophic, change at the end of the Cretaceous. *Nature, Lond.* **292**, 529–31.

Hill, R. R., Jr (1971). Selection in autotetraploids. *Theor. Appl. Genet.* **41**, 181–6.

Hiraizumi, Y. (1971). Spontaneous recombination in *Drosophila melanogaster* males. *Proc. Natl Acad. Sci., USA* **68**, 268–70.

Hirsch, H. R. (1980). Evolution of senescence: influence of age dependent death rates on the natural increase of a hypothetical population. *J. Theor. Biol.* **86**, 149–68.

Ho, M. W. and Saunders, P. T. (1979). Beyond neo-Darwinism – an epigenetic approach to evolution. *J. Theor. Biol.* **78**, 573–91.

Hogenboom, N. G. (1972a). Breaking breeding barriers in *Lycopersicon*. 4. Breakdown of unilateral incompatibility between *L. peruvianum* (L.) Mill. and *L. esculentum* Mill. *Euphytica* **21**, 397–404.

Hogenboom, N. G. (1972b). Breaking breeding barriers in *Lycopersicon*. 5. The inheritance of the unilateral incompatibility between *L. peruvianum* (L.) Mill. and *L. esculentum* Mill. and the genetics of its breakdown. *Euphytica* **21**, 405–14.

Högstedt, G. (1980). Evolution of clutch size in birds: adaptive variation in relation to territory quality. *Science* **210**, 1148–50.

Högstedt, G. (1981). Should there be a positive or negative correlation between survival of adults in a bird population and their clutch size? *Am. Nat.* **118**, 568–71.

Holmquist, R. (1972). Empirical support for a stochastic model of evolution. *J. Mol. Evol.* **1**, 211–22.

Holmquist, R. and Cimino, J. B. (1980). A general method for biological inference: illustrated by the estimation of gene nucleotide transition probabilities. *Biosystems* **12**, 1–22.

Holmquist, R. and Pearl, D. (1980). Theoretical foundations for quantitative paleogenetics. III. The molecular divergence of nucleic acids and proteins for the case of genetic events of unequal probability. *J. Mol. Evol.* **16**, 211–67.

Honne, B. I. (1979). Movements towards equilibrium in autotetraploid populations. *Hereditas* **90**, 85–92.

Hosemann, H. (1948). Schwangerschaftsdauer und Neugeborenegewicht. *Arch. Gynaek.* **176**, 109–176.

Huskins, C. L. (1930). The origin of *Spartina townsendii*. *Genetica* **12**, 531–8.

Hutchinson, G. E. (1959). Homage to Santa Rosalia *or* why are there so many kinds of animals? *Am. Nat.* **93**, 145–59.

Huxley, J. S. (1932). *Problems of Relative Growth*. Methuen, London.

Huxley, T. H. (1887*a*). Scientific and pseudo-scientific realism. *Nineteenth Century* February.

Huxley, T. H. (1887*b*). Science and pseudo-science. *Nineteenth Century* April.

Huxley, T. H. (1889). Agnosticism. *Nineteenth Century* February.

Huxley, T. H. (1895). *Collected Essays XI. Evolution and Ethics and Other Essays*. Macmillan, London.

Imai, H. T. and Crozier, R. H. (1980). Quantitative analysis of directionality in mammalian karyotype evolution. *Am. Nat.* **116**, 537–69.

Jacob, F. (1970). *La Logique du Vivant*. Gallimard, Paris.

Jacobs, P. A. (1981). Mutation rates of structural gene rearrangements in man. *Am. J. Hum. Genet.* **33**, 44–54.

James, S. H. (1970). A demonstration of a possible mechanism of sympatric divergence using simulation techniques. *Heredity* **25**, 241–52.

Jantz, R. L. and Webb, R. S. (1980). Dermatoglyphic asymmetry as a measure of canalization. *Ann. Hum. Biol.* **7**, 489–93.

Jeffrey, E. C. (1915). Some fundamental objections to the mutation theory of de Vries. *Am. Nat.* **49**, 5–21.

Jensen, R. J. and Barbour, C. D. (1981). A phylogenetic reconstruction of the Mexican Cyprinid genus *Algansea*. *Syst. Zool.* **30**, 41–57.

Johnson, M. S. and Mickevich, M. F. (1977). Variability and evolutionary rates of characters. *Evolution* **31**, 642–8.

Joynt, R. J. (1974). The corpus callosum: history of thought regarding its function. In *Hemispheric Disconnection and Cerebral Function* (eds M. Kinsbourne and W. L. Smith), pp. 117–25. C. C. Thomas, Springfield, Illinois.

Jukes, T. H. (1980). Silent evolutionary substitutions and the molecular evolutionary clock. *Science* **210**, 973–8.

Kacser, H. and Burns, J. A. (1981). The molecular basis of dominance. *Genetics* **97** 639–66.

Kalmus, H. (1965). *Diagnosis and Genetics of Defective Colour Vision*. Pergamon Press, Oxford.

Kaneshiro, K. Y. (1976). Ethological isolation and phylogeny in the planitibia subgroup of Hawaiian *Drosophila*. *Am. Nat.* **111**, 897–902.

Kaneshiro, K. Y. (1980). Sexual isolation, speciation and the direction of evolution. *Evolution* **34**, 437–44.

Karn, M. N. and Penrose, L. S. (1951). Birth weight and gestation time in regulation to maternal age, parity and infant survival. *Ann. Eugen.* **16**, 147–64.

Kempthorne, O. (1957). *An Introduction to Genetic Statistics*. Wiley, New York.

Kettlewell, H. B. D. (1973). *The Evolution of Melanism*. Clarendon Press, Oxford.

Kidwell, M. G., Kidwell, J. F. and Sved, J. A. (1977). Hybrid dysgenesis in *Drosophila melanogaster*: a syndrome of aberrant traits including mutation, sterility and male recombination. *Genetics* **86**, 813–33.

Kimura, M. (1961). Natural selection as the process of accumulating genetic information in adaptive evolution. *Genet. Res.* **2**, 127–40.

Kimura, M. (1968). Evolutionary rate at the molecular level. *Nature, Lond.* **217**, 624–6.

Kimura, M. (1977). The neutral theory of molecular evolution and polymorphism. *Scientia* **112**, 687–707.

Kimura, M. (1980*a*). Average time until fixation of a mutant allele in a finite population under continued mutation pressure: studies by analytical, numerical, and pseudo-sampling method. *Proc. Natl Acad. Sci., USA* **77**, 522–6.

Kimura, M. (1980*b*). A simple method for estimating evolutionary rates of base substitutions through comparative studies of nucleotide sequences. *J. Mol. Evol.* **16**, 111–20.

Kimura, M. (1981*a*). Doubt about studies of globin evolution based on maximum parsimony codons and the augmentation procedure. *J. Mol. Evol.* **17**, 121–22.

Kimura, M. (1981*b*). Was globin evolution very rapid in its early stages?: a dubious case against the rate-constancy hypothesis. *J. Mol. Evol.* **17**, 110–13.

Kimura, M. and Ohta, T. (1971*a*). On the rate of molecular evolution. *J. Mol. Evol.* **1**, 1–17.

Kimura, M. and Ohta, T. (1971*b*). *Theoretical Aspects of Population Genetics*. Princeton University Press.

Kimura, M. and Weiss, G. H. (1964). The stepping stone model of population structure and the decrease of genetic correlation with distance. *Genetics* **49**, 561–76.

Kinsbourne, M. and Smith, W. L. (eds) (1974). *Hemispheric Disconnection and Cerebral Function*. C. C. Thomas, Springfield, Illinois.

Kluge, A. G. and Kerfoot, W. C. (1973). The predictability and regularity of character divergence. *Am. Nat.* **107**, 426–43.

Knox, R. B., Willing, R. R. and Pryor, L. D. (1972). Interspecific hybridization in poplars using recognition pollen. *Silvae Genetica* **21**, 65–148.

Kojima, K.-I. (1967). Likelihood of establishing newly induced inversion chromosomes in small populations *Ciencia e Cultura* **19**, 67–77.

Kojima, K.-I. and Tobari, Y. N. (1969). Selective modes associated with karyotypes in *Drosophila ananassae*. II. Heterosis and frequency-dependent selection. *Genetics* **63**, 639–51.

Korey, K. A. (1981). Species number, generation length, and the molecular clock. *Evolution* **35**, 139–47.

Kuchowicz, B. (1971). Diminishing gravitation – a hitherto underrated factor in the evolution of organic life. *Experientia* **26**, 616.

Lack, D. (1947). The significance of clutch size. *Ibis* **89**, 302–52.

Lande, R. (1976). Natural selection and random genetic drift in phenotypic evolution. *Evolution* **30**, 314–34.

Lande, R. (1977*a*). Statistical tests for natural selection of quantitative characters. *Evolution* **31**, 442–4.

Lande, R. (1977*b*). On comparing coefficients of variation. *Syst. Zool.* **26**, 214–7.

Lande, R. (1977*c*). The influence of the mating system on the maintenance of genetic variability in polygenic characters. *Genetics* **86**, 485–98.

Lande, R. (1979). Quantitative genetic analysis of multivariate evolution, applied to brain: body size allometry. *Evolution* **33**, 402–16.

Lande, R. (1980*a*). The genetic covariance between characters maintained by pleiotropic mutations. *Genetics* **94**, 203–15.

Lande, R. (1980*b*). Genetic variation and phenotypic evolution during allopatric speciation. *Am. Nat.* **116**, 463–79.

Lande, R. (1981). Models of speciation by sexual selection in polygenic traits. *Proc. Natl Acad. Sci., USA* **78**, 3721–5.

Lashley, K. S. (1950). In search of the engram. *Symp. Soc. exp. Biol.* **4**, 454–83. Reprinted in F. A. Beach, C. T. Morgan, N. W. Nissen and D. O. Hebb (eds) (1960). *The Neuropsychology of Lashley*. McGraw-Hill, New York.

Lasker, G. W. and Thomas, R. (1976). The relationship between reproductive fitness and anthropometric dimensions in a Mexican population. *Hum. Biol.* **48**, 775.

Lawson, D. A. (1975). Pterosaur from the latest Cretaceous of West Texas: discovery of the largest flying creature. *Science* **187**, 947–8.

Layzer, D. (1978). A macroscopic approach to population genetics. *J. Theor. Biol.* **73**, 769–88.

Layzer, D. (1980). Genetic variation and progressive evolution. *Am. Nat.* **115**, 809–26.

Leps, W. T., Brill, W. J. and Bingham, E. T. (1980). Effect of alfalfa ploidy on nitrogen fixation. *Crop Sci.* **20**, 427–30.

Lessios, H. A. (1979). Use of Panamanian sea urchins to test the molecular clock. *Nature, Lond.* **280**, 599–601.

Lester, L. J. and Selander, R. K. (1981). Genetic relatedness and the social organisation of *Polistes* colones. *Am. Nat.* **117**, 147–66.

Lewis, D. (1960). Genetic control of specificity and activity of the *S* antigen in plants. *Proc. Roy. Soc. Lond.*, *B* **151**, 468–77.

Lewis, D. (1965). A protein dimer hypothesis on incompatibility. *Proc. 11th Int. Congr. Genet.* **3**, 657–63.

Lewis, R. W. (1980). Evolution: a system of theories. *Persp. Biol. Med.* **23**, 551–72.

Lewontin, R. C. (1969). The bases of conflict in biological explanation. *J. Hist. Biol.* **2**, 34–45.

Lewontin, R. C. (1974). *The Genetic Basis of Evolutionary Change*. Columbia University Press, New York.

Lewontin, R. C. (1978*a*). Fitness survival and optimality. In *Analysis of Ecological Systems* (eds D. H. Horn, R. Mitchell and G. R. Stairs). Ohio State University, Columbus.

Lewontin, R. C. (1978*b*). Adaptation. *Sci. Am.* **239**, 212–30.

Lewontin, R. C. and Waddington, C. H. (1967). In *Towards a Theoretical Biology* (ed. C. H. Waddington), Vol. 1. University Press, Edinburgh.

Li, W. H., Gojobori, T. and Nei, M. (1981). Pseudogenes as a paradigm of neutral evolution. *Nature, Lond.* **292**, 237–9.

Liebhaber, S. A., Goossens, M. and Yuet Wai Kan (1981). Homology and concerted evolution at the $\alpha 1$ and $\alpha 2$ loci of human α-globin. *Nature, Lond.* **290**, 26–9.

Lints, F. A. (1978). *Genetics and Aging*. S. Karger, Basel.

Locket, N. A. (1980). Some advances in coelocanth biology. *Proc. Roy. Soc., Lond., B* **208**, 265–307.

Lohrmann, R. and Orgel, L. E. (1979). Self-condensation of activated dinucleotides on polynucleotide templates with alternating sequences. *J. Mol. Evol.* **14**, 243–50.

Lokki, J., Suomalainen, E., Saura, A. and Lankinen, P. (1975). Genetics polymorphism and evolution in parthenogenetic animals. II. Diploid and polyploid *Solenobia Triquetrella* (Lepidoptera: Psychidae). *Genetics* **79**, 513–25.

Long, C. A. (1969). On the use of constancy in estimating conservatism of characters. *Evolution* **23**, 516–7.

Løvtrup, S. (1975). On the falsifiability of neo-Darwinism. *Evol. Theory* **1**, 267–83.

Lovejoy, C. O. (1981). The origin of man. *Science* **211**, 341–50.

Lowe, D. R. (1980). Stromatolites 3,400-Myr old from the Archean of Western Australia. *Nature, Lond.* **284**, 441–3.

Lunn, A. and Haldane, J. B. S. (1935). *Science and the Supernatural: a correspondence between Arnold Lunn and J. B. S. Haldane.* Eyre and Spottiswoode, London.

MacArthur, R. H. and Wilson, E. O. (1963). An equilibrium theory of insular zoogeography. *Evolution* **17**, 373–87.

MacArthur, R. H. and Wilson, E. O. (1967). *The Theory of Island Biogeography.* Princeton University Press.

McClendon, J. H. (1980). The necessary distinction between metabolism and evolution: A comment on the thermodynamics of evolution. *J. Theor. Biol.* **83**, 523–4.

McLaren, A., Chandler, P., Buehr, M., Fierz, W. and Simpson, E. (1981). Immune reactivity of progeny of tetraparental male mice. *Nature, Lond.* **290**, 513–4.

McLean, G. L. (1972). Clutch size and evolution in the Charadrii. *Auk* **89**, 299–324.

McMaster, J. H. (1976). Dynamic analysis of *Pteranodon ingens*: a reptilian adaptation to flight. *J. Paleont.* **49**, 899.

McMorris, F. R. and Zaslavsky, T. (1981). The number of cladistic characters. *Math. Biosciences* **54**, 3–10.

Mark, G. A. and Flessa, K. W. (1977). A test for evolutionary equilibria: phanerozoic brachiopods and cenozoic mammals. *Paleobiology* **3**, 17–22.

Markert, C. L., Shaklee, J. B. and Whitt, G. S. (1975). Evolution of a gene. *Science* **189**, 102–14.

Markow, T. W. (1981). Mating preferences are not predictive of the direction of evolution in experimental populations of drosophila. *Science* **213**, 1405–7.

Marshall, D. R. and Brown, A. H. D. (1981). The evolution of apomixis. *Heredity* **47**, 1–15.

Martin, P. G. and Hayman, D. L. (1966). A complex sex-chromosome system in the hare-wallaby *Lagorchestes conspicillatus* Gould. *Chromosoma* **19**, 159–75.

Maruyama, T. and Imai, H. T. (1981). Evolutionary rate of the mammalian karotype. *J. Theor. Biol.* **90**, 111–21.

Mather, K. F. (1949). *Biometrical Genetics.* Methuen, London.

Matthews, B. W., Grütter, M. G., Anderson, W. F. and Remington, S. J. (1981). Common precursor of lysozymes of hen egg-white and bacteriophage T4. *Nature, Lond.* **290**, 334–5.

Matthey, R. (1973). The chromosome formulae of eutherian mammals. In *Cytotaxonomy and Vertebrate Evolution* (eds A. B. Chiarelli and E. Capanna) pp. 531–616. Academic Press, London and New York.

Maynard Smith, J. (1964). Group selection and kin selection. *Nature, Lond.* **210**, 1145–7.

Maynard Smith, J. (1966). Sympatric speciation. *Am. Nat.* **100**, 637–50.

Maynard Smith, J. (1968). *Mathematical Ideas in Biology.* Cambridge University Press, Cambridge and London.

Maynard Smith, J. (1970). Natural selection and the concept of a protein space. *Nature, Lond.* **255**, 563–4.

Maynard Smith, J. (1971). What use is sex? *J. Theor. Biol.* **30**, 319–35.

Maynard Smith, J. (1978). Optimization theory in evolution. *Ann. Rev. Ecol. Syst.* **9**, 31–56.

Maynard Smith, J. (1980*a*). Models of the evolution of altruism. *Theor. Pop. Biol.* **18**, 151–9.

Maynard Smith, J. (1980*b*). Polymorphism in a varied environment: how robust are the models? *Genet. Res.* **35**, 45–57.

Mayo, O. (1967). Some properties of random in time mating. *Biometrics* **24**, 213.

Mayo, O. (1970). The fixation of new mutants. *Nature, Lond.* **227**, 860.

Mayo, O. (1971). Rates of change in gene frequency in tetrasomic organisms. *Genetica* **42**, 329–37.

Mayo, O. (1973). The evolution of an age limit. *Genetics* **74**, s177.

Mayo, O. (1978) The existence and stability of a three-locus gametophytically-determined self-incompatibility system. *Adv. Appl. Prob.* **10**, 14–15.

Mayo, O. (1980). Variance in clutch size. *Experientia* **36**, 1061–2.

Mayo, O. (1981). The sheltering of lethals on metatherian X-chromosomes. *Genetica* **55**, 27–31.

Mayo, O. and Bishop, J. A. (1979). Comments on S. Løvtrup's paper "On the falsifiability on neo-Darwinism." *Evol. Theory.* **4**, 147–55.

Mayo, O. and Hancock, T. W. (1981). Fixation of genes having large or small effects on a trait with an intermediate optimum. *Hum. Hered.* **31**, 286–290.

Mayo, O., Bishop, G. R. and Hancock, T. W. (1977). Heritability as an indicator of genetical variation in fecundity. *Experientia* **33**, 1024.

Mayo, O., Bishop, G. R. and Hancock, T. W. (1978). The detection of genetical influences of human family size. *Hum. Hered.* **28**, 270–9.

Mayo, O., Hancock, T. W. and Baghurst, P. A. (1980). Influence of major genes on variance within sibships for a quantitative trait. *Ann. Hum. Genet.* **43**, 419–21.

Mayo, O., Eckert, S. R. and Waego Hadi Nugroho (1983). Models of gene effects for a quantitative trait in man. *Proc. 50th Jubilee Conf.* Indian Statistical Institute.

Mayr, E. (1963). *Animal Species and Evolution.* Belknap Press, Harvard.

Mayr, E. (1969). *Principles of Systematic Zoology.* McGraw-Hill, New York.

Mayr, E. (1974). Cladistic analysis or cladistic classification? *Z. Zool. Syst. Evol. Forsch.* **12**, 94–128.

Medawar, P. B. (1952). *An Unsolved Problem of Biology.* H. K. Lewis, London.

Mendel, G. (1865). *Versuche über Pflanzenhybriden.* (English translation ed. J. H. Bennett (1965). Oliver and Boyd, Edinburgh.)

Meyer, J. (1981). A quantitative comparison of the parts of the brains of two Australian marsupials and some eutherian mammals. *Brain Behav. Evol.* **18**, 60–71.

Michod, R. E. (1979). Genetical aspects of kin selection: effects of inbreeding. *J. Theor. Biol.* **81**, 223–33.

Mill, J. S. (1879). *A System of Logic* Book III, Chap. 3, Section 1. London.

Miller, O. J., Sanger, R. and Siniscalco, M. (1978). Report of the committee on the genetic constitution of the X and Y chromosomes. *Cytogenet. Cell Genet.* **22**, 124–8.

Millis, J. and Seng, Y. P. (1954). The effect of age and parity of the mother on birth weight of the offspring. *Ann. Eugen.* **19**, 58–73.

Milton, J. (1671). *Samson Agonistes.* London.

Minvielle, F. (1980). A simulation study of truncation selection for a quantitative trait opposed by natural selection. *Genetics* **94**, 989–1000.

Mitton, J. B. (1975). Fertility differentials in modern societies resulting in normalizing selection for height. *Hum. Biol.* **47**, 189.

Mitton, J. B. (1978). Relationship between heterozygosity for enzyme loci and variation of morphological characters in natural populations. *Nature, Lond.* **273**, 661–2.

Molnar, R. E. and Thulborn, R. A. (1980). First pterosaur from Australia. *Nature, Lond.* **288**, 361–3.

Morton, N. E. (1955). The inheritance of human birth weight. *Ann. Hum. Genet.* **20**, 125–34.

Mosig, G. (1960). Zur Genetik von *Petunia hybrida.* I. Die Selbsterilität. *Z. Vererbungsl.* **91**, 158–63.

Mueller, W. H., Lasker, G. W. and Evans, F. G. (1981). Anthropometric measurements and Darwinian fitness. *J. Biosoc. Sci.* **13**, 309–16.

Mukai, T., Tachida, H. and Ichinose, M. (1980). Selection for variability at loci controlling protein polymorphisms in *Drosophila melanogaster* is very weak at most. *Proc. Natl Acad. Sci., USA* **77**, 4857–60.

Muller, H. J. (1914). A gene for the fourth chromosome of *Drosophila. J. Exp. Zool.* **17**, 325–56.

Muller, H. J. (1925). Why polyploidy is rarer in animals than in plants. *Am. Nat.* **59**, 346–53.

Muller, H. J. (1932). Some genetic aspects of sex. *Am. Nat.* **66**, 118–38.

Muller, H. J. (1942). Isolating mechanisms, evolution and temperature. *Biol. Symp.* **6**, 71–125. (ed. Th. Dobzhansky) Jacques Cattell Press, Lancaster, PA.

Murray, J. and Clarke, B. (1980). The genus *Partula* on Moorea: speciation in progress. *Proc. R. Soc. Lond., B* **211**, 83–117.

Murtagh, C. E. (1977). A unique cytogenetic system in monotremes. *Chromosoma* **63**, 37–57.

Nadeau, J. H., Dietz, K. and Tamarin, R. H. (1981). Gametic selection and the selection component analysis. *Genet. Res.* **37**, 275–84.

Nagel, E. (1979). *Teleology Revisited and Other Essays in the Philosophy and History of Science.* Columbia University Press, New York.

Nebert, D. W. (1981). Possible clinical importance of genetic differences in drug metabolism. *Br. Med. J.* **283**, 537–42.

Neel, J. V. and Rothman, E. (1981). Is there a difference among human populations in the rate with which mutation produces electrophoretic variants? *Proc. Natl Acad. Sci., USA* **78**, 3108–112.

Nei, M. (1969). Gene duplication and nucleotide substitution in evolution. *Nature, Lond.* **221**, 40–2.

Nei, M. (1975). *Molecular Population Genetics and Evolution.* North Holland, Amsterdam.

Nei, M. (1976). Mathematical models of speciation and genetic distance. In *Population Genetics and Ecology* (eds S. Karlin and E. Nevo) pp. 723–65. Academic Press, London and New York.

Nei, M. (1977). Standard error of immunological dating of evolutionary time. *J. Mol. Evol.* **9**, 203–11.

Nelsestuen, G. L. (1980). Origin of life: consideration of alternatives to proteins and nucleic acids. *J. Mol. Evol.* **15**, 59–72.

Nethersole-Thompson, D. (1973). *The Dotterel*. Collins, London.

Nicholas, F. W. and Robertson, A. (1980). The conflict between natural and artificial selection in finite populations. *Theor. Appl. Genet.* **56**, 57–64.

O'Donald, P. (1980). *Genetical Models of Sexual Selection*. Cambridge University Press, London and Cambridge.

Ohno, S. (1969). Evolution of sex chromosomes in mammals. *Ann. Rev. Genet.* **3**, 495–524.

Ohno, S. (1970). *Evolution by Gene Duplication*. Springer-Verlag, New York.

Ohno, S. (1973). Ancient linkage groups and frozen accidents. *Nature, Lond.* **244**, 259–62.

Ohta, T. (1973). Slightly deleterious mutant substitutions in evolution. *Nature, Lond.* **246**, 96–8.

Ohta, T. (1974). Mutational pressure as the main cause of molecular evolution and polymorphism. *Nature, Lond.* **252**, 351–4.

Ohta, T. and Kimura, M. (1971). Functional organization of genetic material as a product of molecular evolution. *Nature, Lond.* **233**, 118–9.

Ohta, T. and Kimura, M. (1981). Some calculations on the amount of selfish DNA. *Proc. Natl Acad. Sci., USA* **78**, 1129–32.

Oka, H.-I. (1957). Genic analysis for the sterility of hybrids between distantly related varieties of cultivated rice. *J. Genetics* **55**, 397–409.

Oka, H.-I. (1974). Analysis of genes controlling the sterility in rice by the use of isogenic lines. *Genetics* **77**, 521–34.

Oliver, A. J., King, D. R. and Mead, R. J. (1979). Fluoroacetate tolerance, a genetic marker in some Australian mammals. *Aust. J. Zool.* **27**, 363–72.

Orgel, L. E. and Crick, F. H. C. (1980). Selfish DNA: the ultimate parasite. *Nature, Lond.* **284**, 604–7.

Orians, G. H. and Pearson, N. E. (1978). On the theory of central place foraging. In *Analysis of Ecological Systems* (eds D. H. Horn, R. Mitchell and G. R. Stairs). Ohio State University, Columbus.

Ornduff, R. (1975). Complementary roles of halictids and syrphids in the pollination of *Jepsonia heterandra* (Saxifragaceae). *Evolution* **29**, 371–3.

Ornston, L. N. and Yeh, W. K. (1979). Origins of metabolic diversity: evolutionary divergence by sequence repetition. *Proc. Natl Acad. Sci., USA* **76**, 3996–4000.

Osborn, H. F. (1909). Darwin and paleontology. In *Fifty years of Darwinism* pp. 209–50. Henry Holt, New York.

Osborn, H. F. (1915). Origin of single characters as observed in fossil and living animals and plants. *Am. Nat.* **49**, 193–239.

Osborn, H. F. (1921). Adaptive radiation and classification of the Proboscidae. *Proc. Natl Acad. Sci., USA* **7**, 231–4.

Osborn, H. F. (1922). Orthogenesis as observed from palaeontological evidence beginning in the year 1889. *Am. Nat.* **56**, 134–43.

Owen, D. R. (1977). In *Evolutionary Ecology* (eds B. Stonehouse and C. M. Perrins) p. 171. Macmillan, London.

Pandey, K. K. (1970). Elements of the *S*-gene complex. VI. Mutations of the self-incompatibility gene, pseudo-compatibility and origin of new self-incompatibility alleles. *Genetica* **41**, 477–516.

Papentin, F. (1980). On order and complexity. I. General consideration. *J. Theor. Biol.* **87**, 421–56.

Patton, J. C., Baker, R. J. and Genoways, H. H. (1980). Apparent chromosomal heterosis in a fossorial mammal. *Am. Nat.* **116**, 143–6.

Pearson, O. P. (1948). Metabolism of small mammals, with remarks on the lower limit of mammalian size. *Science* **108**, 44.

Pearson, P. L. and Roderick, T. H., with Davisson, M. T., Garver, J. J., Warburton, D., Lalley, P. A. and O'Brien, S. J. (1979). *Cytogenet. Cell Genet.* **25**, 82–95.

Pearson, R. (1978). *Climate and Evolution*. Academic Press, London and New York.

Peetz, E. W. (1979). Investigation of charge changes in protein evolution. M.Sc. Thesis, University of California, Berkeley.

Pennycuick, C. J. (1972). *Animal Flight*. Edward Arnold, London.

Perrins, C. M. and Jones, P. J. (1974). The inheritance of clutch size in the Great Tit (*Parus major* L.). *Condor* **76**, 225–9.

Petit, C., Kitagawa, O., Takanashi, E. and Nouaud, D. (1980). The failure to obtain sexual isolation by artificial selection. *Genetica* **54**, 213–9.

Phillips, P. R. and Mayo, O. (1981). Problems in statistical studies on protein polymorphism in natural populations. *Genetics* **97**, 495.

Piazza, A., Menozzi, P. and Cavalli-Sforza, L. L. (1981). Synthetic gene frequency maps of man and selective effects of climate. *Proc. Natl Acad. Sci., USA* **78**, 2638–42.

Picard, G. and L'Héritier, P. (1971). A maternally inherited factor inducing sterility in *Drosophila melanogaster*. *Dros. Inf. Serv.* **46**, 54.

Pilbeam, D. (1980). Major trends in human evolution. In *Current Argument on Early Man* (ed. L. K. Königsson) pp. 261–85. Pergamon Press, Oxford.

Platnick, N. I. (1979). Philosophy and the transformation of cladistics. *Syst. Zool.* **28**, 537–46.

Popper, K. (1957). *The Poverty of Historicism*. Routledge and Kegan Paul, London.

Popper, K. (1974). Darwinism as a metaphysical research programme. In *The philosophy of Karl Popper* Vol. 1. (ed. P. A. Schilpp) pp. 133–43. Open Court, La Salle, IL.

Post, R. H. (1971). Possible cases of relaxed selection in civilized populations. *Humangentik* **13**, 253–84.

Prager, E. M. and Wilson, A. C. (1975). Slow evolutionary loss of the potential for inter-specific hybridization in birds: a manifestation of slow regulatory evolution. *Proc. Natl Acad. Sci., USA* **72**, 200–4.

Pyke, G. H. (1981). Optimal travel speeds of animals. *Am. Nat.* **118**, 475–87.

Ralin, D. and Selander, R. K. (1979). Evolutionary genetics of diploid-tetraploid species of treefrogs of the genus *Hyla*. *Evolution* **33**, 595–608.

Rashin, A. A. (1981). Location of domains in globular proteins. *Nature, Lond.* **291**, 85–7.

Rasmuson, M. (1980). Natural selection and genetic drift in early man. In *Current Argument in Early Man* (ed. L. K. Königsson) pp. 180–1. Pergamon Press, Oxford.

Raup, D. M. (1972). Taxonomic diversity during the phanerozoic. *Science* **177**, 1065–71.

Reed, E. S. (1981). The lawfulness of natural selection. *Am. Nat.* **118**, 61–71.

Rees, H. and Hutchinson, J. (1973). Nuclear DNA variation due to B chromosomes. *Cold Spring Harbor Symp. Quant. Biol.* **38**, 175–82.

Rendel, J. M. (1965). Bristle pattern in *Scute* stocks of *Drosophila melanogaster*. *Am. Nat.* **99**, 25–32.

Rendel, J. M. (1967). *Canalisation and Gene Control*. Logos Press, London.

Rendel, J. M. and Sheldon, B. (1960). Selection for canalization of the scute phenotype in *D. melanogaster*. *Aust. J. Biol. Sci.* **13**, 36–47.

Ricklefs, R. E. and Cox, G. W. (1972). Taxon cycles in the West Indian avifauna. *Am. Nat.* **106**, 195–219.

Riley, R. (1960). The diploidisation of polyploid wheat. *Heredity* **15**, 407–29.

Riska, B. (1979). Character variability and evolutionary rate in *Menidia*. *Evolution* **33**, 1001–4.

Roberts, A. and Tregonning, K. (1980). The robustness of natural systems. *Nature, Lond.* **288**, 265–6.

Robertson, A. (1956). The effect of selection against extreme deviants based on deviation or on homozygosis. *J. Genet.* **54**, 236–48.

Robertson, A. (1968). The spectrum of genetic variation. In *Population Biology and Evolution* (ed. R. C. Lewontin) pp. 5–16. Syracuse University Press, Syracuse, NY.

Robertson, C. (1892). Flowers and Insects IX. *Bot. Gaz.* **17**, 269–76.

Robinson, D. F. and Foulds, L. R. (1981). Comparison of phylogenetic trees. *Math. Biosciences* **53**, 131–47.

Robson, E. B. (1955). Birth weight in cousins. *Ann. Hum. Genet.* **19**, 262–8.

Roff, D. A. (1981). On being the right size. *Am. Nat.* **118**, 405–22.

Rose, M. D. (1976). Bipedal behaviour of olive baboons (*Papio anubis*) and its relevance to an understanding of the evolution of human bipedalism. *Am. J. Phys. Anthropol.* **44**, 247–62.

Rose, M. R. and Charlesworth, B. (1980). A test of evolutionary theories of senescence. *Nature, Lond.* **287**, 141–2.

Rose, M. R. and Charlesworth, B. (1981*a*). Genetics of life history in *Drosophila melanogaster*. I. Sib analysis of adult females. *Genetics* **97**, 173–86.

Rose, M. R. and Charlesworth, B. (1981*b*). Genetics of life history in *Drosophila melanogaster*. II. Exploratory selection experiments. *Genetics* **97**, 187–96.

Rosenzweig, M. L. (1977). On interpreting the results of perturbation experiments performed by nature. *Paleobiology* **3**, 322–24.

Rössler, O. E. (1979). Recursive evolution. *Biosystems* **11**, 193–9.

Ruelle, D. (1981). A mechanism for speciation based on the theory of phase transition. *Math. Biosciences* **56**, 71–5.

Ruse, M. (1977). Karl Popper's philosophy of biology. *Phil. Sci.* **44**, 638–61.

Rutledge, R. W., Basore, B. L. and Mulholland, R. J. (1976). Ecological stability: an information theory viewpoint. *J. Theor. Biol.* **57**, 355–71.

Salisbury, F. B. (1969). Natural selection and the complexity of the gene. *Nature, Lond.* **224**, 342–3.

Samuelson, P. A. (1978). Generalizing Fisher's "reproductive value": overlapping and nonoverlapping generations with competing genotypes. *Proc. Natl Acad. Sci., USA* **75**, 4062–6.

Saunders, P. T. and Ho, M. W. (1981). On the increase in complexity in evolution. *J. Theor. Biol.* **63**, 375–84.

Saunders, P. T. and Ho, M. W. (1981). On the increase in complexity in evolution. II. the relativity of complexity and the principle of minimum increase. *J. Theor. Biol.* **90**, 515–30.

Sawyer, S. and Hartl, D. (1981). On the evolution of behavioural reproductive isolation: the Wallace effect. *Theor. Pop. Biol.* **19**, 261–73.

Schankler, D. M. (1981). Local extinction and ecological re-entry of early Eocene mammals. *Nature, Lond.* **293**, 135–8.

Schindewolf, O. H. (1945). Darwinismus oder Typostrophismus? *Magyar Biol. Kut. Munk.* **16**, 104–77.

Schindewolf, O. H. (1950). *Grundfragen der Paläontologie* p. 495. Schweizerbart, Stuttgard.

Schmalhausen, I. I. (1949). *Factors of Evolution: the Theory of Stabilising Selection.* Blakiston, Philadelphia.

Searcy, W. A. (1980). Optimum body sizes at different ambient temperatures: an energetics explanation of Bergmann's rule. *J. Theor. Biol.* **83**, 579–93.

Seeley, H. G. (1870). *The Ornithosauria: an Elementary Study of the Bones of Pterodactyls.* Cambridge University Press, Cambridge and London.

Selander, R. K., Yang, S. Y., Lewontin, R. C. and Johnston, W. E. (1970). Genetic variation in the horseshoe crab (*Limulus polyphemus*), a phylogenetic "relic". *Evolution* **24**, 402–17.

Sewertzoff, A. N. (1931). *Morphologische Gesetzmässigkeiten der Evolution.* Gustav Fischer, Jena.

Sheppard, P. M. (1969). Evolutionary genetics of animal populations: the study of natural populations. *Proc. XII Inter. Congr. Genet.* **3**, 261–79.

Siegel, A. F. and Fitch, W. M. (1980). Degeneracy when DNA codes for overlapping genes. *Math. Biosciences* **49**, 1–16.

Silberglied, R. E., Aiello, A. and Windsor, D. M. (1980). Disruptive coloration in butterflies: lack of support in *Anartia fatima, Science* **209**, 617–9.

Simpson, G. G. (1949). Rates of evolution in animals. In *Genetics, Paleontology and Evolution* (eds G. L. Jepson, E. Mary and G. G. Simpson) pp. 205–28. Princeton, U.S.

Simpson, G. G. (1953*a*). *The Major Features of Evolution* p. 434. Columbia University Press, New York.

Simpson, G. G. (1953*b*). *Life of the Past.* Yale University Press.

Slatkin, M. (1972). On treating the chromosome as the unit of selection. *Genetics* **72**, 157–68.

Slatkin, M. (1978). On the equilibration of fitnesses by natural selection. *Am. Nat.* **112**, 845–59.

Smit, J. and Hertogen, J. (1980). An extraterrestrial event at the Cretaceous–Tertiary boundary. *Nature, Lond.* **285**, 198–200.

Smith, R. J. (1980). Rethinking allometry. *J. Theor. Biol.* **87**, 97–111.

Smith, R. N. (1981). Inability of tolerant males to sire tolerant progeny. *Nature, Lond.* **292**, 767–8.

Smythies, J. R. (1970). *Brain Mechanisms and Behaviour.* Blackwell Scientific, Oxford.

Snyder, L. R. G. (1980). Evolutionary conservation of linkage groups: additional evidence from murid and cricetid rodents. *Biochem. Genet.* **18**, 209–20.

Sorsby, A. and Fraser, G. R. (1964). Statistical note on the components of ocular refraction in twins. *J. Med. Genet.* **1**, 47–49.

Soulé, M. (1976). Allozyme variation: its determinants in space and time. In *Molecular Evolution* (ed. F. J. Ayala) pp. 60–77. Sinauer Assoc. Inc., Mass.

Southern, E. (1970). Base sequence and evolution of guinea-pig α-satellite DNA. *Nature, Lond.* **227**, 794–8.

Sparrow, A. H. and Nauman, A. F. (1976). Evolution of genome size by DNA doublings. *Science* **192**, 524–9.

Spengler, O. (1919). *The Decline of the West*. London.

Spiess, E. B. (1970). Mating propensity and its genetic basis in *Drosophila*. In *Essays in Evolution and Genetics in honor of Theodosius Dobzhansky* (eds M. K. Hecht and W. C. Steere) pp. 315–79. Appleton-Century–Crofts, New York.

Stanley, S. M. (1975). A theory of evolution above the species level. *Proc. Natl Acad. Sci., USA* **72**, 646–50.

Stebbins, G. L. (1949). The evolutionary significance of natural and artificial polyploids in the family G ramineae. *Proc. VIII Intl Congr. Genet. Hereditas* suppl. 461–85.

Stebbins, G. L. (1971). *Chromosomal Evolution in Flowering Plants*. Addison–Wesley, Reading, Mass.

Stebbins, G. L. and Ayala, F. J. (1981). Is a new evolutionary synthesis necessary? *Science* **213**, 967–71.

Stein, R. S. (1975). Dynamic analysis of *Pteranodon ingens*: a reptilian adaptation to flight. *J. Paleont.* **49**, 534–431.

Stephens, S. G. (1946). The genetics of "corky". I. The New World alleles and their possible role as an interspecific isolating mechanism. *J. Genetics* **47**, 150–61.

Stephens, S. G. (1951). Possible significance of duplication in evolution. *Adv. Genet.* **4**, 247–65.

Stern, C. (1929). Über die additive Wirkung multipler Allele. *Biol. Zentr.* **49**, 261–93.

Stewart, F. M. and Levin, B. R. (1973). Partitioning of resources and the outcome of inter-specific competition: a model and some general considerations. *Am. Nat.* **107**, 171–98.

Strickland, H. E. and Melville, A. G. (1848). *The Dodo and its Kindred*. Benham and Reeve, London.

Strong, D. R. Jr and Simberloff, D. S. (1981). Straining at gnats and swallowing ratios: character displacement. *Evolution* **35**, 810–12.

Sved, J. A. (1968). Possible rates of gene substitution in evolution. *Am. Nat.* **102**, 283–93.

Sved, J. A. (1978). Male recombination in dysgenic hybrids of *Drosophila melanogaster*: chromosome breakage or mitotic crossing-over? *Aust. J. Biol. Sci.* **31**, 303–9.

Sved, J. A. (1979). The "hybrid dysgenesis" syndrome in *Drosophila melanogaster*. *Bioscience* **29**, 659–64.

Sved, J. A. (1981a). A two-sex polygenic model for the evolution of premating isolation. I. Deterministic theory for natural populations. *Genetics* **97**, 197–215.

Sved, J. A. (1981b). A two-sex polygenic model for the evolution of premating isolation. II. Computer simulation of experimental selection procedures. *Genetics* **97**, 217–35.

Sved, J. A. and Mayo, O. (1970). The evolution of dominance. In *Biomathematics*, Vol. 1, *Mathematical Topics in Population Genetics* (ed. K. Kojima) pp. 289–316. Springer-Verlag, Berlin.

Symon, D. E. (1980). The food plants of Australian butterfly larvae. *J. Adel. Bot. Gard.* **2**, 277–92.

Takahata, N. and Maruyama, T. (1981). A mathematical model of extranuclear genes and the genetic variability maintained in a finite population. *Genet. Res.* **37**, 291–302.

Tegelström, H. and Ryttman, H. (1981). Chromosomes in birds (Aves): evolutionary implications of macro- and microchromosome numbers and lengths. *Hereditas* **94**, 225–33.

Templeton, A. R. (1980*a*). Review of *Macroevolution* (S. M. Stanley). *Evolution* **34**, 1224–7.

Templeton, A. R. (1980*b*). The theory of speciation via the founder principle. *Genetics* **94**, 1011–38.

Templeton, A. R. (1981). Some comments on "Genetic variation and progressive evolution" by D. Layzer. *Am. Nat.* **17**, 1049–105.

Tennyson, A. (1833). To J. S. In *Poems* (1842). London.

Terrenato, L., Gravina, M. F., San Martini, A. and Ulizzi, L. (1981). Natural selection associated with birthweight. III. Changes over the last twenty years. *Ann. Hum. Genet.* **45**, 267–78.

Theodoridis, G. C. and Stark, L. (1969). Information as a quantitative criterion of biospheric evolution. *Nature, Lond.* **224**, 860–3.

Theodoridis, G. C. and Stark, L. (1971). Central role of solar information flow in pregenetic evolution. *J. Theor. Biol.* **31**, 377–88.

Thimann, K. V. (ed.) (1980). *Senescence in Plants*. CRC Press, Boca Raton, Florida.

Thoday, J. M. (1958). Homoeostasis in a selection experiment. *Heredity* **12**, 401–15.

Thom, R. (1975). *Structural Stability and Morphogenesis*. Benjamin, New York.

Thompson, D'A. W. (1917). *On Growth and Form*. Cambridge University Press, Cambridge and London.

Thompson, D'A. W. (1942). *On Growth and Form* (2nd edn). Cambridge University Press, Cambridge and London.

Thompson, S. D., Macmillen, R. E., Burke, E. M. and Taylor, C. R. (1980). The energetic cost of bipedal hopping in small mammals. *Nature, Lond.* **287**, 223–4.

Thrailkill, K. M., Birky, C. W. Jr, Lückemann, G. and Wolf, K. (1980). Intracellular population genetics: evidence for random drift of mitochondrial allele frequencies in *Saccharomyces cerevisiae* and *Schizosaccharomyces pombe*. *Genetics* **96**, 237–62.

Tickell, W. L. N. and Pinder, R. (1966). Two-egg clutches in Albatrosses. *Ibis* **108**, 126–9.

Todd, N. B. (1967). A theory of karyotypic fissioning, genetic potentiation and eutherian evolution. *Mamm. Chromsomes Newsl.* **8**, 268–79.

Tôsić, M. and Ayala, F. J. (1980). "Overcompensation" at an enzyme locus in *Drosophila pseudoobscura*. *Genet. Res.* **36**, 57–67.

Tregonning, K. and Roberts, A. (1979). Complex systems which evolve towards homeostasis. *Nature, Lond.* **281**, 563–4.

Turner, J. R. G., Clarke, C. A. and Sheppard, P. M. (1961). Genetics of a difference in the male genitalia of East and West African stocks of *Papilio dardanus* (Lep.). *Nature, Lond.* **191**, 935–6.

Ulizzi, L., Gravina, M. F. and Terrenato, L. (1981). Natural selection associated with birth weight. II. Stabilizing and directional selection. *Ann. Hum. Genet.* **45**, 207–12.

Valentine, J. W. (1969). Patterns of taxonomic and ecological structure of shelf benthos during phanerozoic time. *Paleontology* **12**, 684–709.

Van den Berg, A. and Beintema, J. J. (1975). Non-constant evolution rates in amino acid sequences of guinea pig, chinchilla and coypu pancreatic ribonucleases. *Nature, Lond.* **253**, 207–10.

Van't Hof, J. (1965). Relationships between mitotic cycle time duration, S period duration and the average rate of DNA synthesis in the root-tip meristem cells of several plants. *Exp. Cell Res.* **39**, 48–58.

Van't Hof, J. and Sparrow, A. H. (1963). A relationship between DNA content, nuclear volume, and minimum mitotic cycle time. *Proc. Natl Acad. Sci., USA* **49**, 897–902.

Van Valen, L. (1974). Molecular evolution as predicted by natural selection. *J. Mol. Evol.* **3**, 89–101.

Van Valen, L. (1978). The beginning of the age of mammals. *Evol. Theory* **4**, 45–80.

Van Valen, L. M. and Maiorana, V. C. (1980). The Archaebacteria and eukaryotic origins. *Nature, Lond.* **287**, 248–9.

Van Valen, L. and Sloan, R. E. (1977). Ecology and the extinction of dinosaurs. *Evol. Theory* **2**, 37–64.

Visconti, N. and Delbrück, M. (1953). The mechanism of genetic recombination in phage. *Genetics* **38**, 5–33.

Volterra, V. (1928). Variations and fluctuations of the number of individuals in animal species living together. *J. Cons. Cons. Int. Explor.* **3**, 3–51.

Waddington, C. H. (1953). Genetic assimilation of an acquired character. *Evolution* **7**, 118–26.

Waddington, C. H. (1957). *The Strategy of the Genes.* Allen and Unwin, London.

Waddington, C. H. (1969). In *Beyond Reductionism* (eds A. Koestler and W. H. Thorpe). Hutchinson, London.

Waddington, C. H. and Lewontin, R. C. (1967). A note on evolution and changes in the quantity of genetic information. In *Towards a Theoretical Biology* Vol. 1 (ed. C. H. Waddington) Edinburgh University Press, Edinburgh.

Wade, M. J. (1980a). An experimental study of kin selection. *Evolution* **34**, 844–55.

Wade, M. H. (1980b). Kin selection: its components. *Science* **210**, 665–7.

Wade, M. J. and Breden, F. (1980). The evolution of cheating and selfish behaviour. *Behav. Ecol. Sociobiol.* **7**, 167–72.

Wagner, H. R. (1957). Variation in clutch size at different latitudes. *Auk* **74**, 243–50.

Walker, C. A. (1981). New subclass of birds from the Cretaceous of South America. *Nature, Lond.* **292**, 51–3.

Wallace, A. R. (1889). *Darwinism.* Macmillan, London.

Wallace, B. (1950). An experiment on sexual isolation. *Dros. Inform. Serv.* **24**, 94–6.

Walter, M. R., Buick, R. and Dunlop, J. S. R. (1980). Stromatolites 3,400–3,500 Myr old from the North Pole area, Western Australia. *Nature, Lond.* **284**, 443–5.

Wasserman, G. D. (1978). Testability of the role of natural selection within theories of population genetics and evolution. *Br. J. Phil. Sci.* **29**, 223–42.

Wasserman, M. and Koepfer, H. R. (1977). Character displacement for sexual isolation between *Drosophila mojavensis and Drosophila arizonensis. Evolution* **31**, 812–23.

Wasserman, M. and Koepfer, H. R. (1980). Does asymmetrical mating preference show the direction of evolution? *Evolution* **34**, 1116–24.

Watanabe, R. K. (1979). A gene that rescues the lethal hybrids between *Drosophila melanogaster* and *D. simulans. Jap. J. Genet.* **54**, 325–31.

references   131

Watanabe, T. K. and Kawanishi, M. (1979). Mating preference and the direction of evolution in *Drosophila*. *Science* **205**, 906–7.

Watson, D. M. S. (1974). Pterodactyls past and present. *Phil. Trans. R. Soc. Lond.*, B **267**, 583–5.

Weitkamp, L. R. and Allen, P. Z. (1979). Evolutionary conservation of equine *Gc* alleles and of mammalian *Gc/albumin* linkage. *Genetics* **92**, 1347–54.

West, M. J., King, A. P. and Eastzer, D. H. (1981). The cowbird: reflections on development from an unlikely source. *Am. Scientist* **69**, 56–66.

White, D. H. (1980). A theory for the origin of a self-replicating chemical system. 1. Natural selection of the autogen from short, random oligomers. *J. Mol. Evol.* **16**, 121–47.

White, M. J. D. (1961). Cytogenetics of the grasshopper *Moraba scurra*. *Aust. J. Zool.* **9**, 784–90.

White, M. J. D. (1973). *Animal Cytology and Evolution* 3rd edn. Cambridge University Press, Cambridge and London.

White, M. J. D. (1978). *Modes of Speciation*. Freeman, San Francisco.

Wicken, J. S. (1980). A thermodynamic theory of evolution. *J. Theor. Biol.* **87**, 9–23.

Wiedmann, J. (1969). The heteromorphs and ammonoid extinction. *Biol. Rev.* **44**, 563–602.

Williams, G. C. (1957). Pleiotropy, natural selection, and the evolution of senescence. *Evolution* **11**, 398–411.

Williams, G. C. (1981). *Sex and Evolution*. Princeton University Press, Princeton.

Williams, W. H. (1978). How bad can "good" data really be? *Am. Stat.* **32**, 61–5.

Williams, W. and Gale, J. S. (1960). Effect of selection on the frequency of genotypes determining incompatibility in *Prunus avium*. *Nature, Lond.* **185**, 944–5.

Wilson, A. C., Carlson, S. S. and White, T. J. (1977). Biochemical evolution. *Ann. Rev. Biochem.* **46**, 573–639.

Winkler, H. (1916). Über die experimentelle Erzeugung von Pflanzer mit abweichen Chromosomenzahlen. *Z. Bot.* **8**, 471–53.

Wood Jones, F. and Porteus, S. D. (1928). *The Matrix of the Mind*. University Press Association, Honolulu.

Woodwell, G. M. and Smith, H. H. (eds) (1969). *Brookhaven Symp. Biol.* **22**.

Wool, D., Bunting, S. and Van Emden, H. F. (1978). Electrophoretic study of genetic variation in British *Myzus persicae* (Sulz.) (Hemiptera, Aphididae). *Biochem Genet.* **16**, 987–1006.

Wright, S. (1929). Fisher's theory of dominance. *Am. Nat.* **63**, 274.

Wright, S. (1931). Evolution in Mendelian populations. *Genetics* **16**, 97–159.

Wright, S. (1941). On the probability of fixation of reciprocal translocations. *Am. Nat.* **75**, 513–22.

Wright, S. (1949). Population structure in evolution. *Proc. Am. Phil. Soc.* **93**, 471–8.

Wright, S. (1967). Comments on the preliminary working papers of Eden and Waddington. In *Mathematical Challenges to the neo-Darwinian Interpretation of Evolution*. (eds P. S. Moorehead and M. M. Kaplan). pp. 117–20. Wistar Institute Press, Philadelphia.

Wright, S. and Dobzhansky, Th. (1946). Genetics of natural populations. XII. Experimental reproduction of some of the changes caused by natural selection in certain populations of *Drosophila Pseudoobscurra*. *Genetics* **31**, 125–56.

Wriston, J. C. (1981). Biochemical peculiarities of the guinea pig and some possible examples of convergent evolution. *J. Mol. Evol.* **17**, 1–9.

Yates, F. (1950). The place of statistics in the study of growth and form. *Proc. Roy. Soc. Lond., B* **137**, 479–88.

Yockey, H. P. (1974). An application of information theory to the central dogma and the sequence hypothesis. *J. Theor. Biol.* **46**, 369–406.

Yockey, H. P. (1978). Can the central dogma be derived from information theory? *J. Theor. Biol.* **74**, 149–52.

Yockey, H. P. (1979). Do overlapping genes violate molecular biology and the theory of evolution? *J. Theor. Biol.* **80**, 21–6.

Yokoyama, S. and Felsenstein, J. (1978). A model of kin selection for an altruistic trait considered as a quantitative character. *Proc. Natl Acad. Sci., USA* **75**, 420–2.

Youden, W. J. (1972). Enduring values. *Technometrics* **14**, 1–10.

Young, E. C. (1961). Degeneration of flight-musculature in the corixidae and notonectidae. *Nature, Lond.* **189**, 328–9.

Young, E. C. (1965). Flight muscle polymorphism in British corixidae: ecological observations. *J. Animal Ecol.* **34**, 253–390.

Zeuner, F. E. (1946). *Dating the Past.* Methuen, London.

Zeven, A. C. (1979). Polyploidy and domestication: the origin and survival of polyploids in cytotype mixtures. In *Polyploidy Biological Relevance* (ed. W. H. Lewis) pp. 385–407. Plenum, New York.

Zuckerkandl, E. and Pauling, L. (1965). Evolutionary divergence and convergence in proteins. In *Evolving Genes and Proteins* (eds V. Bryson and H. J. Vogel) pp. 97–166. Academic Press, New York and London.

author index

Abbie, A. A., 8
Ahearn, J. N., 81
Aiello, A., 33
Alberch, P., 19, 104
Alexander, R. D., 73
Allen, P. Z., 57
Allendorf, F. W., 61
Allison, A. C., 36
Alvarez, L. W., 99
Alvarez, W., 99
Anderson, E., 87, 96
Anderson, W. F., 9, 47
Antonovics, J., 80
Apple, M. S., 18
Aquadro, C. F., 45
Arbous, A. G., 64
Archibald, J. D., 98
Armstrong, R. A., 90
Arnheim, N., 25
Arnold, J. T. A., 55
Asaro, F., 99
Astolfi, P., 89
Atchley, W. R., 19
Avise, J. C., 45
Axelrod, R., 73
Ayala, F. J., 36, 90

Baba, M. L., 44
Bader, R. S., 51
Baghurst, P. A., 51, 60
Bailey, G. S., 61
Bailey, S. M., 104
Baker, R. J., 52–4
Balch, W. E., 4
Baldwin, J. E., 31
Ball, J. A., 18
Barbieri, M., 24
Barbour, C. D., 89
Barlow, H. B., 28

Basore, B. L., 15
Bateman, A. J., 83–5
Bawa, K. S., 83
Beach, J. H., 83
Beckman, G., 67
Beckner, M., 1
Beintema, J. J., 45
Bell, G., 22
Bellingham, A. J., 47
Bender, E. A., 77
Bengtsson, B. O., 52–4, 57–8
Bennett, J. H., 26, 66, 76
Bennett, M. D., 91
Bergson, H., 4
Bickham, J. W., 53–4
Bingham, E. T., 91
Birky, C. W., Jr, 26
Bishop, G. R., 37, 51
Bishop, J. A., 36, 97
Blakemore, R., 4
Blaustein, A. R., 72
Blundell, T., 47
Boaz, N. T., 105
Bodmer, W. F., 52, 57–8
Bonen, L., 4
Bookchin, R. M., 47
Bookstein, F. L., 96
Boucot, A. J., 98
Boulter, D., 44
Boyd, R., 73
Boyer, S. H., 46–7
Bradshaw, A. D., 3
Brady, R. H., 1, 8, 29, 36
Bramwell, C. D., 20, 48
Brandham, P. E., 56–7
Breden, F., 73
Brent, L., 6
Bridges, C. B., 61
Brill, W. J., 91

Britten, R. J., 61
Brooks, L. D., 78
Brown, A. H. D., 83
Brown, W. L., Jr, 81
Brown, W. M., 45
Bryant, P. J., 81
Buehr, M., 6
Buick, R., 11
Bull, C. M., 87
Bunting, S., 92
Burke, E. M., 22
Burla, H., 52
Burley, N., 62
Burns, J. A., 67, 104
Bush, G. L., 54
Buss, L. W., 76

Cachel, S., 102–3
Calow, P., 65
Campbell, R. B., 79
Carlson, S. S., 44–5
Carr, B. J., 22
Carson, H. L., 81, 95
Case, S. M., 54
Case, T. J., 77
Cavalli-Sforza, L. L., 73, 74, 81, 103
Cavener, D. R., 66
Cedérgren, R. J., 25
Chakraborty, R., 43
Chandler, P., 6
Chapman, R. A., 77
Charlesworth, B., 57, 76, 85–6, 88
Charlesworth, D., 57, 85–6
Charnov, E. L., 62
Chen, K. N., 4
Chew, G. K., 92
Chomsky, N., 102
Christiansen, F. B., 78
Cifelli, R. A., 101
Cimino, J. B., 11
Clarke, B., 94
Clarke, C. A., 6, 82, 89
Clarke, P. H., 61
Clegg, M. T., 66
Clemens, W., 98
Cloud, P., 11
Clutton-Brock, T. H., 19
Cody, M. L., 62
Colgan, D. J., 36
Colwell, R. K., 78

Connor, E. F., 16
Cooper, D. W., 56, 92
Corbin, K. W., 46
Cox, G. W., 101
Cracraft, J., 95
Crick, F. H. C., 25
Cronin, J. E., 105
Crosby, J. L., 83
Crow, J. F., 36, 37, 39
Crozier, R. H., 53
Cruden, R. W., 80
Cuénot, L., 5
Cullis, C. A., 6
Czelusniak, J., 44

Darga, L. L., 44
Darlington, C. D., 14, 91
Darlington, P. J., 74
Darwin, C. R., 2, 5, 22, 24, 60, 70, 78, 86, 89, 95, 102
Davenport, C. B., 5
Davidson, E. H., 61
Dawkins, R., 74
de Jong, W. W., 44, 46
Delbrück, M., 26
Demetrius, L., 39
de Vries, H., 5
de Wet, J. M. J., 93
Diamond, J. M., 97
Dickinson, H., 80
Dietz, K., 71
Dirac, P. A. M., 14
Dobzhansky, Th., 57, 78, 86, 87, 89
Dodson, M. M., 59, 89
Doolittle, W. F., 25
Duncan, C. J., 32–3
Dunlop, J. S. R., 11
Durrant, A., 6
Dyer, T. A., 4

Eanes, W. F., 60
Eastzer, D. H., 69, 82
Eberhard, W. G., 16
Economos, A. C., 17
Edson, M. M., 105
Edwardson, J. R., 87
Eigen, M., 24, 26
Eigenmann, C. H., 27
Eldredge, N., 54, 95–6
Evans, F. G., 104
Ewens, W. J., 38, 41, 46

Falconer, D. S., 39, 73
Farris, J. S., 51
Feldman, M., 61
Feldman, M. W., 73–4, 78
Felsenstein, J., 15, 73, 78, 80, 89, 94
Ferris, S. D., 45
Fierz, W., 6
Fisher, R. A., 3, 4, 5, 16, 18, 19, 22,
    27–8, 30–2, 36, 38, 65, 66–7, 70–2.
    75, 77, 78, 83, 94
Fitch, W. M., 44, 45
Flessa, K. W., 101
Foin, T. C., 105
Ford, E. B., 3, 67
Foster, G. G., 55
Foulds, L. R., 89
Fox, G. E., 4
Fraccaro, M., 29
Francke, U., 55
Franklin, I. R., 26
Fraser, G. R., 49
Freeman, D. D., 81
Freire-Maia, N., 70
Freud, S., 103

Gale, J. S., 83
Ganapathy, R., 99
Gans, C., 6
Gardiner, W. C., Jr, 24
Garn, S. M., 104
Gartner, S., 99
Genoways, H. H., 52
Georgeson, M. A., 33
Gibbon, E., 2
Gibson, J., 4
Gillespie, J. H., 40, 62, 79
Gingerich, P. D., 50, 96
Gladyshev, G. P., 26
Godfrey, L., 96
Gojobori, T., 48
Gold, J. R., 54
Goldschmidt, R., 5
Goodman, M., 43–4
Goossens, M., 25, 61
Gorczynski, R. M., 6
Gosse, E., 1–2
Gosse, P., 1–2
Gould, S. J., 19, 54, 60, 90, 95–6, 100,
    104
Grant, P. R., 81

Gravina, M. F., 30
Gray, H., 18
Greaves, J. T., 44
Green, R. F., 33
Greenewalt, C. H., 20–2
Gregory, R. P. G., 3
Grey, D. R., 65
Griffing, J. B., 72
Griffiths, R. C., 79
Grosberg, R. K., 80
Grosjean, H., 25
Grüneberg, H., 59
Grütter, M. G., 11, 47
Gupta, R., 4
Gurwitsch, A., 5
Gustafson, L. B., 11
Guthrie, R. D., 74

Haldane, J. B. S., 14, 17, 29, 34, 36, 41,
    48, 49, 61, 66, 73, 75, 91, 96, 105
Hall, B. G., 61
Hallam, A., 89
Hamilton, W. D., 72–3, 78
Hampé, A., 19
Hanbin, E. H., 20
Hancock, T. W., 37, 51, 62, 66
Handford, P., 60
Hansen, O., 102
Hardin, G., 90
Harper, C. W., Jr, 89
Harper, K. T., 81
Harris, M. G., 33
Harrison, R. G., 59
Hart, G. E., 60
Hartl, D. L., 52, 80
Harvey, P. H., 19
Haukioja, E., 30, 63
Hayman, D. L., 13, 53
Heckel, D. G., 20
Hed, H. M. E., 37
Hedrick, P. W., 55
Hendrickson, J. A., Jr, 81
Henle, P., 105
Hennig, W., 89–91
Herbers, J. M., 29
Herm, D., 98
Hermann-Parker, S. M., 80
Hertogen, J., 98
Hespell, R. B., 4
Hewitt, G. M., 91

Heyde, C. C., 62–3
Hickey, L. J., 99
Hill, R. R., Jr, 91
Hiraizumi, Y., 87
Hirsch, H. R., 75
Ho, M. W., 15, 66
Hogenboom, N. G., 81
Högstedt, G., 30, 69
Holmquist, R., 11, 44
Honne, B. I., 91
Hosemann, H., 29
Huskins, C. L., 92
Hutchinson, G. E., 80
Hutchinson, J., 91
Huxley, J. S., 18
Huxley, T. H., 8, 9, 33, 105

Ichinose, M., 59
Imai, H. T., 53–4

Jacob, F., 6
Jacobs, K. H., 96
Jacobs, P. A., 55
James, S. H., 94
Jantz, R. L., 60
Jeffrey, E. C., 5
Jensen, R. J., 89
Johnson, M. S., 51
Johnston, P. G., 56, 92
Johnston, W. E., 104
Jones, P. J., 62
Joynt, R. J., 6, 8
Jukes, T. H., 44
Jungen, H., 52

Kacser, H., 67, 104
Kalmus, H., 49
Kan, Y. W., 25, 61
Kaneshiro, K. Y., 81–2
Karn, M. N., 29
Kawanishi, M., 81–2
Keany, J., 99
Kempthorne, O., 39
Kerfoot, W. C., 51
Kerrich, J. E., 64
Kettlewell, H. B. D., 36, 67
Khasanov, M. M., 26
Kidd, K. K., 89
Kidwell, J. F., 87
Kidwell, M. G., 87

Kimura, M., 13, 26, 42–5, 48, 60, 79, 104
King, A. P., 69, 82
King, D. R., 9
Kinsbourne, M., 8
Kitagawa, O., 79, 80
Klikoff, L. G., 81
Kluge, A. G., 51, 96
Knapp, C. M., 105
Knox, R. B., 83
Koepfer, H. R., 81
Kojima, K.-I., 57
Konovalov, C., 55
Korey, K. A., 45
Korostyshevskiy, M. A., 18
Krebs, H., 31
Krebs, J. H., 62
Krystal, M., 25
Kuchowicz, B., 14
Kunkel, L. M., 46–7

Lack, D., 62
Lande, R., 40, 49–51, 58, 71, 94
Langston, P. J., 60
Lankinen, P., 92
La Palme, G., 25
La Rue, B., 25
Lashley, K. S., 8
Lasker, G. W., 104
Lawson, D. A., 20–1
Layzer, D., 40, 77, 79
Leps, W. T., 91
Lessios, H. A., 45
Lester, L. J., 74
Levin, B. R., 79
Lewis, B. J., 4
Lewis, D., 85
Lewis, R. W., 105
Lewontin, R. C., 1, 3, 4, 8, 13, 26, 28, 39, 79, 104
L'Héritier, P., 87
Li, W. H., 48
Liebhaber, S. A., 25, 61
Lindley, P., 47
Lints, F. A., 74
Locket, N. A., 48
Lohrmann, R., 25
Lokki, J., 92
Long, C. A., 51
Lovejoy, C. O., 103
Løvtrup, S., 97

Lowe, D. R., 11
Lückemann, G., 26
Luehrsen, K. R., 4
Lunn, A., 105

MacArthur, R. H., 20, 100
McClendon, J. H., 14
Mace, G. M., 19
McGehee, R., 90
McGuirk, J. P., 99
McLaren, A., 6
McLean, G. L., 64
McMaster, J. H., 20
MacMillen, R. E., 22
McMorris, F. R., 89
McNeilly, T. S., 3
Maffi, G., 55
Magrum, L. J., 4
Maiorana, V. C., 4
Maniloff, J., 4
Mark, G. A., 100
Markow, T. W., 81
Marshall, D. R., 83
Martin, P. G., 13, 53
Maruyama, T., 26, 53, 87
Mather, K. F., 26
Matsuda, G., 44
Matthews, B. W., 11, 47
Matthey, R., 53
Maxson, R. D., 45
Maynard Smith, J., 10, 17, 29, 30, 72–4,
    77, 80, 94
Maynes, G. M., 56, 92
Mayo, O., 4, 13, 26, 36, 37, 38, 39, 45,
    51, 53, 55, 60, 62, 64, 66, 67–8, 70,
    75, 79, 83, 91, 97
Mayr, E., 13, 78, 89, 93, 94, 96
Mead, R. J., 9
Medawar, P. B., 6, 75
Melville, A. G., 13
Mendel, G., 4
Menozzi, O., 103
Meyer, J., 7
Michel, H. V., 99
Michod, R. E., 72
Mickevich, M. F., 51
Mill, J. S., 104
Miller, L., 47
Miller, O. J., 55
Millis, J., 29

Milton, J., 34
Mitton, J. B., 60, 104
Minvielle, F., 104
Molnar, R. E., 48
Moore, G. W., 44
Morrison, K., 11
Morton, N. E., 29
Mosig, G., 86
Moss, D., 47
Mueller, W. H., 104
Mukai, T., 58
Mulholland, R. T., 15
Muller, H. J., 61, 77, 81, 86, 91
Murray, J., 94
Murtagh, C. E., 53

Nadeau, J. H., 71
Nagel, E., 2
Nagel, R. L., 47
Nagylaki, T., 39
Nauman, A. F., 93
Nebert, D. W., 62
Neel, J. V., 43
Nei, M., 45, 48, 61, 82
Nelsestuen, G. L., 25
Nethersole-Thompson, D., 63
Nicholas, F. W., 104
Nouaud, D., 79, 80

O'Donald, P., 71
Ohno, S., 53, 55, 61, 69, 91, 93
O'Hara, R. K., 72
Ohta, T., 26, 42–3, 45, 61, 104
Oka, H.-I., 87
Oliver, A. J., 9
Orgel, L. E., 25
Orians, G. H., 28
Ornduff, R., 83
Ornston, L. N., 25
Osborn, H. F., 4, 89, 90
Oster, G. F., 19, 104
Owen, D. F., 62

Pandey, K. K., 85
Papentin, F., 14–15
Patton, J. C., 45, 54
Patton, J. L., 52
Pauling, L., 43
Pearl, D., 44
Pearson, N. E., 28

Pearson, O. P., 17
Pearson, P. L., 55–6
Pearson, R., 98
Peetz, E. W., 47
Pennycuick, C. J., 21
Penrose, L. S., 29
Perrins, C. M., 62
Petit, C., 79, 80
Phillips, P. R., 36
Piazza, A., 103
Picard, G., 87
Pilbeam, D., 103
Pinder, R., 64
Platnick, N. I., 89, 103
Popper, K., 3
Porteus, S. D., 7–8
Post, R. H., 49
Poulter, R. T. M., 61
Prager, E. M., 54
Pyke, G. H., 21

Rak, Y., 105
Ralin, D., 92
Ranney, H. M., 47
Rashin, A. A., 47
Rasmuson, M., 37
Raup, D. M., 100, 101
Rayfield, L. S., 6
Reed, E. S., 1
Rees, H., 91
Rees, M. J., 22
Remington, S. J., 9, 47
Rendel, J. M., 65
Richerson, P. J., 73
Ricklefs, R. E., 101
Riley, R., 93
Riska, B., 51
Roberts, A., 15
Robertson, A., 40, 66, 104
Robertson, C., 82
Robinson, D. F., 89
Robson, E. B., 29, 70
Roderick, T. H., 55–6
Roff, D. A., 20
Rose, M. D., 103
Rose, M. R., 76
Rosenzweig, M. L., 50
Rössler, O. E., 59
Rothman, E., 43
Roughgarden, J., 20

Ruelle, D., 89
Ruse, M., 3
Rutledge, J. J., 19
Rutledge, R. W., 15
Rydén, L, 46
Ryder, O., 25
Ryttman, H., 54

Salisbury, F. B., 10
Samuelson, P. A., 38
San Martini, A., 30
Sanger, R., 56
Sankoff, D., 25
Sapienza, C., 26
Saunders, P. T., 15, 66
Saura, A., 92
Sawyer, S., 80
Schankler, D. M., 50
Scharloo, W., 88
Schindewolf, O. H., 100
Schmalhausen, I. I., 40
Schmickel, R., 25
Schopf, T. J. M., 100
Schuh, H. J., 62
Schuster, P., 24
Scott, A. F., 46–7
Searcy, W. A., 18
Seeley, H. G., 48
Selander, R. K., 74, 92, 104
Seng, Y. P., 29
Sepkoski, J. J., Jr, 100
Sewertzoff, A. N., 40
Sharp, P., 53
Sheldon, B. L., 65
Sheppard, P. M., 6, 33, 82, 89
Siegel, A. F., 44
Silberglied, R. E., 33
Simberloff, D. S., 16, 81
Simons, E. L., 50
Simpson, E., 6
Simpson, G. G., 7, 96, 101
Siniscalco, M., 56
Slatkin, M., 26, 30
Slingsby, C., 47
Sloan, R. E., 97, 99
Smit, J., 98
Smith, H. H., 16
Smith, K. D., 46–7
Smith, R. J., 18
Smith, R. N., 6

Smith, W. L., 8
Smythies, J. R., 7
Snyder, L. R. G., 58
Sorsby, A., 49
Soulé, M., 51
Southern, E., 25
Sparrow, R. H., 91, 93
Spengler, O., 100
Spiess, E. B., 79
Stackebrandt, E., 4
Stahl, D. A., 4
Stanley, S. M., 54
Stark, L., 12–14
Stebbins, G. L., 90, 91, 92, 93, 96
Steele, E. J., 6
Stein, R. S., 20
Stephens, S. G., 61, 83
Stern, C., 67
Stewart, F. M., 79
Stock, C. S., 5
Stockwell, P. A., 61
Strickland, H. E., 13
Stringer, C. B., 105
Strong, D. R., Jr, 81
Soumalainen, E., 92
Sved, J. A., 5, 42, 67, 80, 87–8
Symon, D. E., 95, 98

Tachida, H., 59
Taggart, R. T., 55
Takahata, N., 26, 87
Takanashi, E., 79, 80
Tamarin, R. H., 71
Tanner, R. S., 4
Taylor, C. R., 22
Tegelström, H., 54
Templeton, A. R., 41, 54, 94–5
Tennyson, A., 3
Terrenato, L., 30
Theodoridis, G. C., 12–14
Thimann, K. V., 75
Thoday, J. M., 77
Thom, R., 59
Thomas, R., 104
Thompson, D'A. W., 18
Thompson, S. D., 22
Thomson, G., 40
Thornton, I. W. B., 6
Thrailkill, K. M., 26
Thulborn, R. A., 48

Tickell, W. L. N., 64
Tickle, I., 47
Tobari, Y. N., 57
Todd, N. B., 53
Tôsić, M., 36
Tregonning, K., 15
Turnell, B., 47
Turner, J. R. G., 82

Ulizzi, L., 30
Uzzell, T., 46

Valentine, J. W., 100
van den Berg, A., 45
Vandeberg, J. L., 56, 92
van't Hoff, J., 91
van Emden, H. F., 92
van Noordwijk, A. J., 88
van Valen, L. M., 4, 28, 97, 99
Veitch, C. R., 97
Visconti, N., 26
Volterra, V., 90

Waddington, C. H., 13, 39, 59, 65–6
Wade, M. J., 72–3
Wagner, H. R., 63
Wake, D. B., 19, 104
Walker, C. A., 48
Wallace, A. R., 81
Wallace, B., 79
Walter, M. R., 11
Wasserman, G. D., 3
Wasserman, M., 81
Watanabe, T. K., 81–2, 86
Watson, D. M. S., 20, 48
Watson, J. A. L., 11
Webb, R. S., 60
Weiss, G. H., 79
Weitkamp, L. R., 57
West, M. J., 69, 82
White, D. H., 24–6
White, M. J. D., 54, 55, 81, 86, 91
White, T. J., 44, 54
Whitfield, G. R., 20, 48
Whitten, M. J., 55
Wicken, J. S., 59
Wiedmann, J., 98, 100
Williams, G. D., 75–6, 77
Williams, W., 83
Williams, W. H., 43

Wilson, A. C., 44, 45, 54
Wilson, E. O., 20, 81, 100
Wilson, G., 25
Windsor, D. M., 33
Wistow, G., 47
Woese, C. R., 4
Wolf, K., 26
Wood Jones F., 7–8
Woodwell, G. M., 16
Wool, D., 92
Wright, S., 4, 31, 57, 66, 86, 90, 95
Wriston, J. C., 45

Yang, S. Y., 104

Yates, F., 18
Yeh, W. K., 25
Yockey, H. P., 44
Yokoyama, S., 73, 77
Youden, W. J., 43
Young, E. C., 59

Zablen, L. B., 4
Zaslavsky, T., 89
Zeuner, F. E., 13
Zeven, A. C., 92
Zimmer, E., 25
Zuckerkandl, E., 43

subject index

Acquired characters, inheritance of, 6
Adaptation, 9, 27–33, 53
  defined, 27–8
Agrostis tenuis, 3
Albumin linkage, 57–8
Allometry, 18–19, 50, 104
Allopatric speciation, 94–5
Allopolyploidy, 92–3
Aloe, 57
α-crystallin, 44
α-galactosidase, 55–6
α-globin genes, 25, 61
Aneuploidy, 55–7
Archaeobacteria, 4
Artificial selection, *see* Selection
Asexual reproduction, 77, 93
Asteroid impact and extinction, 99
Asymmetry, 26
  and canalization, 60
Autogen, 25, 26
Autopolyploidy, 92–3

B-chromosome, 93
  accumulation, 91
Bacteriophage T4, 47
Balanced translocation, *see* Translocation
Baluchitherium, 17
β-haemoglobin mutants, 46–8
Bipedality, 103, 105
Bird flight, 20–2
Birthweight, human, 29–30
Biston betularia, 103
  melanism in, 36
Body size and fertility, 20
Brachiosaurus, 17
Brain,
  eutherian, 6–8
  marsupial, 7–8

Brain – *cont.*
  size, 50
  and bodyweight, 19
  hominid, 102–3
Brood parasitism, 69
Buffering, 67
"Burnt fingers" distribution, 64

Canalization, 59–69, 104
  defined, 59
Carp, evolutionary change in, 47
Casuarinaceae, 98
Catastrophe theory, 89
Cavia porcellus, 45
Central dogma, 44
Character displacement, 81, 101
Charadrius
  alexandrinus, 63
  melanops, 63
  melodus, 63
  wilsonia, 63
Chenopodiaceae, 98
Chimpanzee, X chromosomes of, 56
Chromosome evolution, 52–8, 60–2, 90–3
Climatic changes and extinction, 98–9
Clutch size, 30, 62–5
  distribution, 63–4
  heritability, 62
  of shorebirds, 63–4
  variance, 62–4
Coelocanth, stasis and change in, 48, 104
Colonizing species, 95–8
Colour blindness, 49
Coloration, disruptive, 33
Competitive exclusion, 89–90
Complexity,
  biological, 14–16
  ecological, 14–16

Convergent evolution, 45
Co-operation, 72, 76
Cope's rule, 49
Corixids, 59–60
Corpus callosum, 6–8
Correlated selection response, 39–40
Cost of meiosis, 88
Cost of natural selection, 41–2
Cranial nerves, human, 60
Creation, 2, 10
Cretaceous–Tertiary boundary, 98–9
    explanations for extinctions, 98–9
Cricetinae, 45
Cryptodirans, 54
Cyprinidae, 54
Cytochrome *c*, evolutionary change in,
    44, 47
Cytochrome P-450, 62

Daphnia carinata, 92
Dejerine, 8
Deleterious mutants, chance fixation of,
    46–7
Didelphis, 101
Dinosaur evolution, 17, 48–9
DNA, 16, 24–6, 44
    functionless, 25–6
    quantity, evolutionary changes in, 13,
        25, 44, 45, 91, 93
    repair, 46
Dodo, extinction of, 13, 97
Dollo's 'law', 97, 100
Dominance, evolution of, 4, 66–9, 75
Dosage compensation, 42, 60–2, 66
Drift, 35, 43, 49–50, 52, 57, 65–6
Drosophila, 20, 96
Drosophila melanogaster
    deleterious genes in, 58
    genetical assimilation in, 65
    hybrid dysgenesis in, 87–8
    interspecific crossing in, 86–7
    inversion in, 57
    mating preference, 80–2
Drosophila pseudoobscura, 45, 50
    genetical variability in, 92
Drosophila persimilis, genetical
    variability in, 92
Drosophila robusta, genetical variability
    in, 92

Drosophila silvestris, sexual selection in,
    81
Drosophila simulans, interspecific
    crossing in, 86–7
Duplication, 60–2, 93

Emberiza schoeniclus, 30
Emberizidae, 45
Environmental deterioration, 27–8, 34
Enzyme kinetics and dominance, 67
Epiloia, 36
Equilibrium, evolutionary, 34
Equus, 58
Eudromias morinellus, 63
Extinction, 13, 97–101
    patterns in, 99–101
Eyesight, human, 28, 49

Family selection, *see* Selection
Fecundity, 20, 62–5
Fibrinopeptides, evolutionary change in,
    44, 46–7
Fish,
    evolutionary change in, 54
    polyploid, 61
Fission, chromosomal, 53
Fitness, 34–42
    changes in, 35, 38, 42
    inclusive, 72–4
Flight, 20–2, 97
Fluoroacetate poison, 9
Food chains and extinction, 99
Fossil record, 37, 48, 50, 101
Founder effect, 94–5
Fundamental theorem of natural
    selection, 38, 42
    corollary, 39, 49
Fusion, chromosomal, 53

Galen, 8
Gall, 8
Gallus domesticus, 47
Gasteria, 57
Gastrolobium, 9
Gene inactivation, 60–2
Gene regulation in eukaryotes, 61–2
Genetical assimilation, 65
Genetical code, 44
Genetical material, primordial, 24–6
Genetical variability, 51, 57, 92

Genome doubling, 93
Glucose-6-phosphate dehydrogenase, 56
Goodeniaceae, 98
Gorilla, X chromosomes of, 56
Gorilla gorilla, 56
Gravitational constant, 14
Great tit,
    clutch size, 62
    inbreeding, 88
Group selection, *see* Selection
Guinea-pig, ribonucleases in, 45

Haemoglobin,
    α chain, 61
    β chain, 46–7
    evolutionary change in, 44, 46–7
Haemoglobin C, 47
Haemoglobin S, 36, 47
Haemoglobin San Diego, 46–7
Haworthia, 57
Heritability, 19, 40, 50
Heterochrony, 104
Heterozygosity, 51
Heterozygote,
    advantage, 35–6, 79
    variability, 61, 66–9
Histone IV, 44
Hominid evolution, 102–3
Homo sapiens, 56, 58
Homoeostasis, 60–1
Hopping, 22–3
Horse, linkage in, 58
Houstonia purpurea, 82
Huntington's chorea, 75
Hybrid,
    breakdown, 87–8
    dysgenesis, 87–8
    inviability, 86–7
    sterility, 87
    zone, 82, 87
Hyla chrysoscelis, 92
Hyla versicolor, 92
Hypercycle, 24
Hypoxanthine guanine phosphoribosyl
    transferase, 55–6
Hyopsodus, 96
Hystricomorph rodents, ribonucleases
    in, 45

Inbreeding, 31–2, 72–3, 88

Inclusive fitness, *see* Fitness
Incompatibility, *see also* Self-
    incompatibility
    interspecific, 82
Index of opportunity for selection, 37
Information, 12–14, 51
Insulin, evolutionary change in, 44, 47
Inversion, chromosomal, 55, 57
    frequency, 57
Isolation,
    by artificial selection, 79–80
    behavioural, 81–2
    ecological, 80
    gametic, 83–6
    mechanical, 82
    pollination, 82–3
    reproductive, 78–88
    temporal, 80–1
    zygotic, 86–8

K-selection, 20
Kangaroo, X chromosome of, 56
Karyotype,
    change, 52–8
    optimum, 53
Kelvin, Lord, 11
Kin selection, *see* Selection

Lamarck, J. B., 6
Latimeria, 104
Life history, 19–20
Limulus, 96, 101
Linkage, genetical, 55–8, 77–8
Load, substitutional, 41–2
Lobibyx novae-hollandiae, 63
Lolium remotum, 91
Lolium perenne, 91
Lotka–Volterra equation, 15
Lucilia cuprina, 55
Lungfish evolution, 13, 104
Lycopersicon, 82
Lysozyme, evolutionary change in, 44

Macropus eugenii, 9
Macropus giganteus, 56
Magnetic field, 26
Magpie, clutch size of, 30
Male sterility, 82, 87
Man, 102–3
    linkage in, 55–6
    X chromosome of, 56

Manual dexterity, 103
Marsupial,
    extinction, 7, 13
    genetical variability, 92
Meiotic drive, 4
Melanism, industrial, 36–7
Mendelism, 1
Menidia, 51
Messenger RNA, *see* RNA
Mimicry, 5–6, 32–3
Molothrus ater, 69
Moraba scurra, 55
Mouse, X chromosome of, 56, 58
Mus musculus, 56–8
Musca domestica, 55
Mutation, 43–8
    chromosomal, 57
    neutral, 43–4
    in poultry, 5
Myoglobin, evolutionary change in, 44, 47
Myopia, 49
Myrtaceae, 98
Myzus persicae, 92

Natural selection, *see* Selection
Neutrality, selective, 34, 43, 45–8
Nicotiana–type self-incompatibility, 83–4
Normalizing selection, *see* Selection
Notoryctes typhlops, 49

Ocean junction and extinction, 89
Oenothera lamarckiana, balanced translocation in, 5, 77
Onset age in genetical diseases, 75
Ontogeny, 19, 40, 104
Optimal foraging, 29, 33
Orangoutan, X chromosome of, 56
Origin of life, 11, 24
Ornithine transcarbamylase, 56
Orthogenesis, 4, 9
Oxylobium, 9

Pan troglodytes, 56
Panaxia dominula, 3
Papilio dardanus, 82
Papilio memnon, 6
Paralidae, 45
Paratettix texanus, 36

Partula, 94
Parus major, 62, 88
Pelycodus, 50, 96
Peripheral isolates, 95
Peromyscus maniculatus, 58
Petunia hybrida, 86
Pharmacogenetics, 62
Phosphoglycerate kinase, 56
Pica pica, 30
Plasticity, developmental, 59
Plovers, clutch size in, 62–5
Polistes, 74
Polycystic kidney syndrome, 66
Polymophism, balanced, *see*
    Heterozygote advantage
Polyploidy, 39, 47, 90–2, 101
    and genetical variability, 92
    gene frequency change in, 91
Pongo pygmaeus, 56
Power–size relationships, 17
Pre-adaptation, 5–6, 9
Predator–prey interaction, 16
Proteaceae, 98
"Protein clock", 44–5
Protein variability, 10–11, 45–8
"Pseudo-genes", 47–8
Pterosaur flight, 20–1
Pterosaur fossil, 48

Quantitative trait change,
    measured, 48–51
    tabulated, 49
Quetzlcoatl northropi, 20

r-selection, 20
Rana cascadae, 72
Random-in-time mating, 26
Ranidella insignifera, 87
Ranidella pseudoinsignifera, 87
Rat, X chromosome of, 56
Rate of amino acid substitution, 42–8
    estimation, 42–4
Rate of chromosome evolution, 52–8
Rate of evolution, 11, 34–58, 77
Rate of gene substitution, 34–6, 41–2
Rate of nucleotide substitution, 44
Rattus fuscipes, 9
Rattus norvegicus, 56
Recombination, 55–8
    modification, 55, 77–8

Reed bunting clutch size, 30
Reproductive isolation, *see* Isolation
Ring species, 94
Ribonuclease, evolutionary change in,
    44–5
RNA, 24
    messenger, 62
    ribosomal, 24

Sea-urchins, rate of evolution in, 56
Secale cereale, 91
Segregation distortion, 57
Selection,
    artificial, 104
    and canalization, 60, 65–9
    dominance-modifying, 66–9
    family, 73
    group, 72
    individual, 20, 34–7
    intensity, 29–30, 34–7, 58
    kin, 71–4
    mechanisms of, 34–6, 39
    affecting mutation and
        recombination, 40
    natural, 2 *et seq.*
        defined, 2–3
    normalizing, 66
    sexual, 70–1, 81
    single gene, 34–7
    stabilizing, 36, 66
Selenia bilunaria, 36
Self-fertilization, 31–2
Self-incompatibility, 70, 83–6, 88
Senescence, 74–6
    and adaptation, 74–5
    in plants, 75
Sensory perception, 28, 32–3
Serranidae, 78
Sex, 77–88
Sex-chromosomes in metatherians, 56
Sexual dimorphism, 70–1
Sexual reproduction, 77–88
Sexual selection, *see* Selection
Size, 17–18
Skull, human, 60
Social insects, 74, 78
Solenobia triquetrella, 92

Spartina alternifolia, 92
Spartina maritima, 92
Spartina townsendii, 92
Speciation, 78–80, 89–97, 99–101
    allopatric, 94
    pattern, 99–101
    through polyploidy, 90–3
    rates, 95–7
    stimulation, of, 100
    sympatric, 93–4
Spencer, Herbert, 3
Sphenodon, 101
Stabilizing selection, *see* Selection
Stasis, 34, 96, 100, 101, 104
Supergenes, 6
Sympatry, 93–4
Synteny, 55–8
    conservation, 55–8, 60–1

Talpa europea, 49
Taxon cycle, 101
Threshold, 32
Time, 10–12
Translocation, 52
    balanced, 77, 86
Tribolium, 76
Triticum aestivum, 61–2
Typolysis, 100

Umbelliferae, 80

Vanellus tricolor, 63
Variance, components, 38–9
    phenotypic, 50
Vesalius, 8
Vicq d'Azyr, 8
Vitamin D binding protein, linkage,
    57–8

Whale, 54
Wheat, triplicate genes, 61–2
Willis, 8
Williston's rule, 1, 15
Wing evolution, 20–2

X-chromosome stability, 55–7, 69